Designing Mobile
Autonomous Robots

D1707954

Designing Mobile Autonomous Robots

by John Holland

ELSEVIER

AMSTERDAM • BOSTON • HEIDELBERG • LONDON
NEW YORK • OXFORD • PARIS • SAN DIEGO
SAN FRANCISCO • SINGAPORE • SYDNEY • TOKYO

Newnes

Newnes is an imprint of Elsevier
200 Wheeler Road, Burlington, MA 01803, USA
Linacre House, Jordan Hill, Oxford OX2 8DP, UK

 Recognizing the importance of preserving what has been written, Elsevier prints its books on
acid-free paper whenever possible.

Library of Congress Cataloging-in-Publication Data

(Application submitted.)

British Library Cataloguing-in-Publication Data
A catalogue record for this book is available from the British Library.

ISBN: 0-7506-7683-3

For information on all Newnes publications
visit our website at www.newnespress.com

03 04 05 06 07 08 10 9 8 7 6 5 4 3 2 1

Printed in the United States of America

Dedication

This book is dedicated to all the employees, board members, stockholders and supporters of Cybermotion over all the years. How I should ever have been lucky enough to work with so many dedicated, talented, intelligent, humorous and tenacious people I shall never know. All of the challenges we faced together have produced these pages, I merely did the typing. In fairness, it must also be dedicated to the long suffering spouses of the "Cyberdogs," including my own wonderful wife Sheilah.

Contents

Contents

Foreword

One of the most exciting challenges a designer can face is that of building a practical autonomous mobile robot. This is as close as we mere mortals can come to producing a living being. Make no mistake; we are talking about very primitive beings, but artificial beings nonetheless. Autonomous robots exist today that can perform complex tasks without human assistance for weeks or even months, but these robots will seem laughably crude in the years to come.

The building blocks for autonomous robots have been readily available for several years. Powerful microprocessors, laser-based sensors (lidar), Ethernet radio communications, video processors, and a host of other subsystems are now priced at levels that permit practical autonomous machines to be built for an exciting range of commercially viable applications. There are even a wide range of simple sensors and actuators that allow the hobbyist to develop small, but sophisticated robots.

The challenge is in understanding how these systems can be made to play together in a coherent and effective way to create a system that is far more than the sum of its parts. If the designer thinks of a new robot design as being laser-guided, or as using GPS navigation, the result will be a design that is inflexible. Such a design may be useful, but it will be not able to grow beyond its initial concept. A stripe following "Automatic Guided Vehicle" is an excellent example of such a design. Autonomous robots are much more robust and interesting beasts.

It is my experience that any good concept will have an intrinsic elegance. A good software and hardware structure is like a snowflake, with each subsystem having the same basic structure as every other subsystem. At the center, a few basic structures hold it all together. Each point of the snowflake will have differences from the others, but will follow the same basic pattern.

This is not a book full of complex equations. Everything you need to know about math you learned in geometry class. Nor is this a book about how to build a robot. Instead, this is a book about how to organize a robot design so that building it follows naturally. Robot design is not so much about inventing as it is about rediscovering. The concepts are all familiar; they just need to be placed into a snowflake.

The ideas presented here are based on 18 years of robot design experience. There can be no doubt that others have discovered many of these concepts and given them different names. As I have said, this is about discovery as much as it is about invention. Let me apologize in advance if I have failed to give anyone credit here.

Finally, designing an autonomous robot teaches us many priceless lessons, not the least of which is a deep humility and appreciation of the miracle of the simplest living creatures. Yet for me, there is nothing that compares to the thrill of turning your creation loose to fend for itself! Once you have mastered these concepts, you will be able to approach the design of any complex control system with complete confidence.

What's on the CD-ROM?

Included on the accompanying CD-ROM:

- A full searchable eBook version of the text in Adobe pdf format
- A directory containing the sourcecode for all of the example programs in the book

Refer to the ReadMe file for more details on CD-ROM content.

Section 1:
Background Software Concepts

CHAPTER 1

Measure Twice, Cut Once

The chances are pretty good that you wouldn't have picked up this book unless you already had a strong motivation to push the envelope and create a marvelous new machine. You may have already created one or more robots and are anxious to put your experience in perspective and do it all again. You are probably anxious to get on with bending metal and writing code as soon as possible.

As mentioned in the preface, this is not a book about "How to Build" a robot. If it were, you could simply copy some instructions and be building your machine immediately. Instead, this book is about how to organize your approach so that you can begin to create innovative machines that can react to ever-changing conditions to perform useful tasks. The difference is that between being an artist, and being great at paint-by-numbers.

Determinism

And why is designing a mobile robot so much more complex than, say, writing an accounting program? In scale it may not be, but the inputs to most software applications are of finite variety, and calculations are absolute. There is only one correct balance for a column of numbers. Even complex graphics programs have only one set of outputs for any given set of inputs.

With a mobile robot, however, the input combinations change every time it runs a path. Nothing stays constant in the physical world, and nothing looks the same twice in a row. So nearly infinite are the combination of stimuli to which a sensor-based robot may be exposed that its behavior can be described as virtually nondeterministic. In other words, it appears that the exact behavior of the system cannot be predicted from merely observing its operation and the environment. The keyword here is "virtually," since the permutations of stimuli and reactions become too convoled for prediction through even the most enlightened observation.

When we truly begin to grasp this fact, the whole process of programming autonomous systems takes on an almost metaphysical dimension. Input data can no longer be processed at face value, but must be filtered in relation to other data, and according to experiences of the recent past. In the process of doing this, we begin to see our own human behavior in a new light. Fear and caution, for example, begin to look like excellent control modification mechanisms, and not merely human emotions.

Given the complexity of the task ahead, we must plan our approach carefully. The second most frustrating phase of any complex project is determining the architecture to be used. The first most frustrating phase is realizing after months of trying to implement a flawed architecture that it has degenerated into a mass of patches and expedients, that it has lost any semblance of elegance, and that it will have to be razed to the ground and rebuilt.

While developing our architecture, we try to think of all the various capabilities we will need and vaguely how they will be performed in hardware and software, and then envision a structure that can be used to tie them together. We then think of things that could go wrong, and challenge our ephemeral structure. This often causes the rickety conceptual framework to crash to the ground, but each time it does so we build anew with more robust geometries. Often a vicious circle of reasoning will be encountered which repeatedly flips us from one approach to another, only to be pushed back again by complex considerations, but with single-minded persistence our architecture will eventually emerge.

I will warn of as many traps as I can, but others you will have to find yourself. Just remember that you always need a "Plan B." That is, you must always keep in mind how you will adapt your approach if one or more elements prove flawed.

Rule-based systems, state-driven systems, and other potential tar pits

To some extent, rule-based systems and state-driven systems are simply opposite sides of the same coin. In its simplest form, a state-driven system attempts to provide a discrete block of instructions (rules) that define the machine's reaction to its inputs for a given state or circumstance. It is almost inevitable that any architecture will have elements of such a structure, but *appropriately defining the states is the key*.

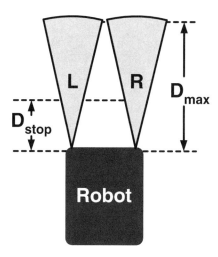

Figure 1.1. Simple robot with two sonar sensors

As an example of how quickly a purely state-driven system can become unwieldy, let's look at a very primitive robot. In our example, the robot has two forward-looking sonar detectors to provide collision avoidance, and it has two modes: "Off" and "Drive." We will assume it can only drive forward. To simplify the issue even more, we will forget about navigation and assume we simply want it to keep moving.

Now let's solve the problem with a discrete, single level, state-driven solution. Single level simply means that there will be no states within states. We determine that there are the following states:

State Number	Mode	L>Dmax and R>Dmax	L<R	R<L	Behavior
1	OFF	NC	NC	NC	Turn motors off
2	FORWARD	X			Accelerate forward
3	FORWARD		x		Turn right while decelerating
4	FORWARD			x	Turn left while decelerating

Table 1.1. Simple states for robot of Figure 1.1

5

Coded as a simple linear program, the program flow would resemble that shown in Figure 1.2. The first task gets the sonar ranges. The robot's mode and the ranges are then tested to determine the state. The State 3 and 4 functions will calculate the robot's rate of turning, and its acceleration or deceleration.

State 1 simply decelerates at the highest safe rate to bring the robot to a halt if it is not already halted. State 2 (all clear) accelerates the robot while steering straight ahead, and states 3 and 4 cause the robot to turn while decelerating. These functions must continuously recalculate the deceleration so that the robot will come to a halt if and when either sonar range becomes less than the programmed stopping distance (Dstop).

The desired result for State 3 and 4 behaviors is for the robot's path to arc at an increasingly tight angle as it approaches a target because the speed is decreasing as the steering rate is increasing. Hopefully, the target will eventually pass away to the side of the sonar transducer pattern, State 2 will become active, and our robot will continue on its way.

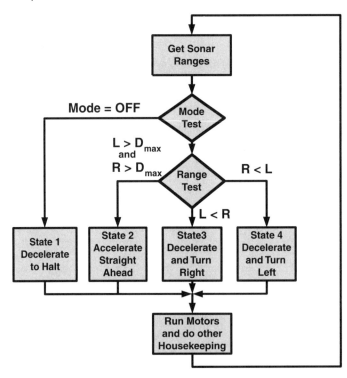

Figure 1.2. Program flow for robot of Figure 1.1

So what could go wrong? The answer is just about everything! The first problem has to do with timing. If the task that gets the sonar ranges is actually running the sonar, or it must wait for the ranges to be determined by another system, then the length of time required to process the range information will vary depending on conditions. Furthermore, the calculations of the various states will take variable amounts of time. The reason that this is important is that the states must calculate accelerations and velocities, and to do this they must be related to a consistent time base.

The final task of running the motors will be executing a control program that will also require consistent timing. It is therefore necessary that the program loop at a constant and known rate so that these calculations can be made. Reading a hardware clock is not an option, as the sampling rate of the system needs to be constant to assure stability. If the reader is familiar with filter theory, all of this will sound very familiar! The simplest solution to the problem of achieving constant timing is to eliminate the loop back to the top of the process, and execute the whole process from a timer interrupt.

If we assume that the process is timer driven, it will be necessary to assure that the task is completed before the timer fires again. If the timer starts again before calculations are complete, then partially calculated parameters may be overwritten and contaminated. A task like this one is thus said to be nonreentrant. Worse yet, repeatedly reentering a nonreentrant task can cause the processor to eventually overflow its stack and crash. This can cause very strange and unwelcome behavior in our little appliance.

Next consider the "straw man" scenario of Figure 1.3. Here our robot encounters a target on its left side (State 3), begins turning right, drives into the path of a target on that side at an even closer range (State 4). As it turns away from the target on the right, it goes back into State 3. The result is that the robot drives forward so that the closer target approaches along the right edge of the right transducer pattern until the robot comes to a halt, or worse, sideswipes the closer target (Figure 1.4).

It has already become apparent that our control architecture is flawed. We must have a strategy for escaping from such a situation. One problem with our strategy is that we are trying to solve the problem instantaneously, without regard to the recent past. As soon as we begin to turn away from the closer target, we forget all about it, and react to the further target again. Truly, a robot that doesn't learn from history is doomed to repeat it!

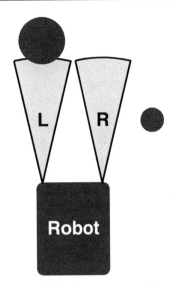
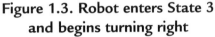

Figure 1.3. Robot enters State 3 and begins turning right

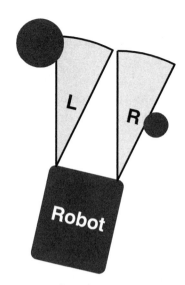

Figure 1.4. Robot becomes trapped

We could elect to go in any of several directions with our simple robot. We could provide latched states to take into account recent history. For example, we could record the fact that the system had gone between states 3 and 4 without ever having experienced a State 2 clear period. If this happened, we could force the robot to continue to turn away from the nearer obstacle until State 2 was achieved.

To do this, the robot would have to essentially latch the behavior of State 4 (turn left) if it occurred immediately after State 3. We could then unlatch State 4 if State 2 (Clear) or State 1 (Halt) conditions occurred. To do this we could introduce new, *persistent* states called states 5 and 6. State 5 could be called "Escape Right" and State 6 could be called "Escape Left." A persistent state remains true unless cancelled by a higher ranking state.

Figure 1.5 shows the logic flow for our simple collision avoidance robot as a modified state-driven machine. Notice that we have gone from an instantaneous state appraisal and reaction to a more complex decision tree in which some states must be evaluated before others, and where the previous state may be maintained even if its initial trigger conditions are no longer true. The logic no longer has the simple elegance promised in the concept of a "state-driven" machine. The flow is already getting convoluted, and we have yet to give the robot any real sense of objective! Worse, there are still potential weaknesses in the performance. For example, there is little chance that we will maneuver between the two obstacles in Figure 1.4, even if the distance between them is ample.

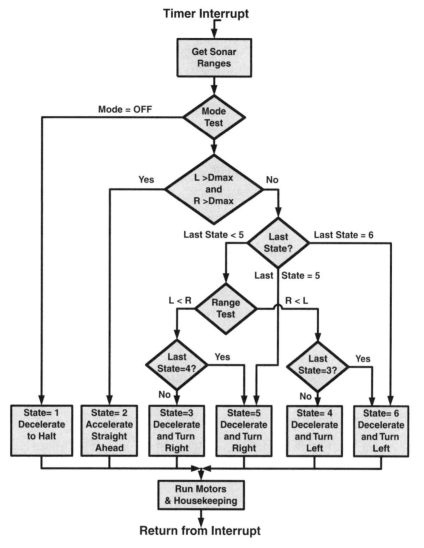

Figure 1.5. State driven program flow with escape logic

When one begins to think of adding rules to help the robot achieve an objective, and to maximize its circumnavigation performance, it becomes obvious that the purely state-driven model will be very messy indeed. What is more, it will become very difficult to maintain and modify the structure.

A proper architecture will have an inherent elegance, just as an efficient airplane is usually aesthetically pleasing to look at. When this elegance is not evident, it is probably not an optimal architecture.

Defining an open architecture

Having spent only a short while with our state-driven model, it quickly became evident that we were going to need to continue to add to the architecture as we tested it against real or hypothetical challenges. It is therefore obvious that our architecture needs to be flexible and modular in nature.

The first rule of creating such an architecture should, therefore, be to think in general terms rather than specific terms. We should think of each element of the structure as a specific case of a generic element, and consider how the element can be made as flexible as possible.

For example, if we are considering elements that control the speed and the rate of turning of the robot in response to sonar data, we should also consider how our system might be expanded to use another and as yet undetermined sensor. We need to consider how this similar module might be used at the same time or in lieu of the sonar sensor interface without disrupting other elements of the control architecture. And we should push the thought further to consider an interface that doesn't report range, such as a bumper contact. How is such an event going to get processed into our robot's behavior?

Both the real sonar module and our hypothetical sensor will need to calculate deceleration and turning, so we would want to break out the functions that calculate these parameters and make them called subroutines (functions).

We might also add a structure that tests all available collision avoidance data to merge the readings into a single assessment of the situation. In the end we would then create a sonar module that plugged into a general framework. If we take the time to do this, even though we can't envision needing another type of sensor, we will be well on our way to developing a living design instead of a rigid, inflexible monument to our lack of foresight.

In the final analysis, an architecture is like a living organism operating under the laws of natural selection. As it is subjected to challenges, it will either grow stronger or be destroyed. In many cases, the architecture itself will not survive, but it will give birth to an offspring that does.

A Brief History of Software Concepts

To say that autonomous vehicle design encompasses a vast array of concepts and disciplines is an understatement. A robot designer needs a practical working knowledge of classical disciplines such as geometry, calculus, statics, dynamics, thermodynamics, and biology. Additionally, a working familiarity with narrower disciplines from battery technology to signal processing is important. Fortunately, it isn't necessary to have the law-of-cosines at one's fingertips, but only to know of its existence and possible usefulness. The World Wide Web has placed information on every topic imaginable at our fingertips. We have only to know to look for it.

Likewise, if one is to effectively program such a system, a wide knowledge of many software concepts is essential. No one concept or language is sufficient for all the requirements, but rather these concepts are the palette from which you will create your masterpiece. The richer the palette, the greater the potential will be. Again, it is only necessary at the start to have an overview of as many tools as possible.

There is a natural tendency for experts in each of these fields to generate a mystique around their specialties. In the course of this book, I hope to demystify many of these concepts, and to help you to grasp the core of their meaning. Not only are most of these concepts relatively simple at their center, but many of the concepts are mirrored across multiple disciplines. Once you have identified a need to incorporate a concept into your design, you can easily find reference material to help you gain the depth you require in the subject.

The more we know about one thing, the more we know about many things. The universe is a huge fractal.

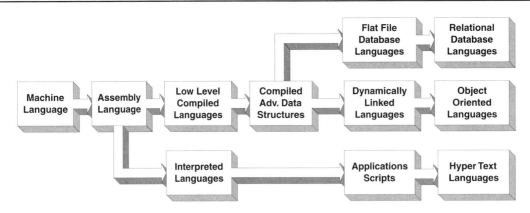

Figure 2.1. A grossly over-simplified history of the evolution of software

The history of software evolution is a useful context for understanding the reason for each concept and its potential usefulness. In the beginning there was machine language. This is the stuff all computers burn, no matter how abstracted it becomes from the programmer. I would argue that it is as useful to understand the basics of Boolean mathematics as it is the more advanced concepts.

Assembly language

It wasn't long before early programmers realized that the first programs they needed to write were tools to facilitate programming itself. Essentially, the first significant market for software was to other programmers. Thus, the first program language to evolve was assembly language, a more-or-less direct representation of the machine language it generated. Simple text editors were created in order to write the assembly language programs. These text files were then *assembled* into binary machine language.

Before long, programmers began to create huge unmanageable assembly language programs, and the concept of modular code emerged. Now programs were written in manageable chunks, which could be *included* (copied by the assembler) into other programs, or which could be assembled into *object* code that could be *linked* to the object code from other modules to produce the final machine language program[1].

Mention assembly language to a freshly minted computer science major and the reception will be similar to that a leper might enjoy on a particularly bad hair day.

---

[1] This use of the word object has no relation to the object-oriented concepts to be discussed later.

Yet no language can be more efficient than assembly language because at some stage all languages must either generate or run on machine code. If you are willing to invest the time, you can gather a library of assembly language routines that elevate application programming to a level similar to a higher level language.

In programming systems on the CyberGuard™ robot, we developed not only an extensive library of powerful assembly language routines, but also an efficient real-time operating kernel and several dozen specialized scripting languages. The geometric functions were all specifically written to use the native units of the robot's encoders. For example, instead of *degrees* these routines expected *begrees.* Begrees are an angular measurement whereby 1024 begrees is a full circle, the output of the robot's angular encoders. Units of measure for vector calculations were all based on 100ths of a foot, the basic odometry scaling. The result was that small, inexpensive, low-power processors could perform calculations as complex as linear regression with blazing speed.

The smaller the microprocessor, the more likely you will need to use assembly language at some point. This is particularly true in programming sensor and control systems for mobile robots because this type of real-time programming is not generally well supported by available languages. As you become more proficient with assembly language, you will naturally begin to incorporate many of the concepts of higher-level languages.

As programs become more elegant, they become fractal in nature. Patterns repeat themselves, mirroring the most powerful constructs of other languages.

The trade-off between assembly language and higher-level languages is of course efficiency versus development time and code maintainability. With processing power as incredibly inexpensive as it is today, only a masochist would program a major application for a PC in assembly language. Yet assembly language may yet be the best (and sometimes only) recourse for programming many subsystems of a mobile robot.

Early conventional languages

As the programs began to diversify into scientific, financial, and engineering applications, early conventional languages evolved to facilitate programming in each application area. Languages such as FORTRAN and LISP evolved for scientific applications while COBOL became dominant in many early finance applications. *Basic* soon appeared primarily as an easy-to-learn program for the non-geek masses. Early versions of *Basic* violated almost every precept of a good programming discipline, but as time went by new incarnations of the language began borrowing the best ideas from other languages.

That humble *Basic* would evolve into the massively powerful programming environment that is VB, and that it would come to displace many of the specialized languages in their own native disciplines is an amazing testament to the power of the concepts it embraces. But we are getting ahead of our little history.

In conventional programming the basic elements are simple:

- Main line code (or threads in the case of multithreaded programs)
- Functions (subroutines)
- Variables

Main line code is the place where the high-level thinking goes on, and the functions are called to solve particular problems. Functions need input to act on, so function calls came to include *arguments* to tell the function what the calling program wanted to do. For example, if a function is to multiply two numbers, then the arguments would naturally be the numbers to be multiplied. Since functions could call other functions, libraries quickly evolved to unwieldy size.

Variables began to appear in an ever greater variety of flavors ranging from Boolean (true/false) and integers to Variants (chameleon-like types that can take on the form of almost any data type) and from single variables to multidimensional arrays and user-defined data types (defined collections of other variables). The more complex structures resulted from more complex programs and more complex data environments.

In the earliest languages, all variables were global—any function could access any variable. This of course led to some very evil things. As computers began using interrupts to service simple I/O requirements, it became almost certain that there would be conflicts. For this reason, the concept of "local" variables quickly took root. A local variable is owned by a function, and is often stored on the processor's stack by the function. If *all* a function's variables are local, then it is said to be reentrant (more will be said about this subject in the next chapter).

At this point in software evolution, many of the qualities of object-oriented programming were already beginning to appear. Once functions could be called with arguments and perform all of their calculations using only local variables, then a function could be called from any number of places, even before it had finished serving other tasks. Each time it was called, it would simply save its work on the stack and start the new problem. Unless the programmer makes specific efforts to the contrary, the work of a function is last-in first-out (LIFO). The function's use of the stack to store data meant that for all intents and purposes the function could appear to clone itself. It was as if each client that called the function owned its own copy. This concept would later evolve into objects that could be *instanced*.

Compilers vs. interpreters

Early conventional languages were either *compiled* or *interpreted*. A compiler reads the text of the program and generates the final machine language it represents. The result is a stand-alone piece of code that the computer can execute. The compiler is not needed by the user of the code.

An interpreter, on the other hand, resides in the end user's computer and reads and executes the text when the application is running. Thus the computer is executing the interpreter, and the interpreter is executing the program. A hybrid of these concepts is a pseudo compiler that generates a compressed *pseudocode* that an interpreter then reads and executes at run time. In efficiency, pseudocode systems are somewhere between strict interpreters and true compilers.

Compiled programs run far faster than interpreted programs, and for some time it appeared that interpreted programs would go the way of the Dodo bird. This was not to be. Instead, specialized interpreters began to appear that allowed the user to write a simple *script* to customize the operation of an application.

As the Internet became more popular and important, scripting languages such as HTML and JAVA script evolved to provide a simple way of generating interactive web pages.

Scripted languages are most appropriate when a single instruction generates a great deal of machine work. When this is true, the inherent inefficiency of interpreting is negated.

The robots we developed at Cybermotion use a scripted language called the *SGV or Self-Guided Vehicle* language. This language has slightly over 100 instructions, and it is used to generate a simple *pseudocode* that is interpreted by the robot at run time. Thus, to make a robot run from one part of a building to another, one has only to transmit a few hundred bytes of this pseudocode to the robot. The vast bulk of the required code, including the interpreter, is permanently burned into the memory of the onboard processors.

```
;//////////////////////////////////////////////////////////////////
;        B13_ABAP.sgv- Path file generated by PathCAD Version 3.63.
;
;        Path program from B13_AB to B13_AP.
;        Map Name: B13 [B13]
;        Map Desc: B13 Warehouse
;
;        Generated by: JMH
;        Generated on: 01/16/02 at: 08:54
;//////////////////////////////////////////////////////////////////
;        --- INCLUDE FILES ---
;
        INCLUDE local.def      ;Local operating parameter definitions.
        INCLUDE global.def     ;System constants and addresses.
        INCLUDE B13.def        ;Map definitions from GENDEF.
;
;------------------------------------------------------------------
;
        AVOID   NC,     Far,Wide           ;Set sonar, front=Far, side=Wide.
        FID     1,      TG_7               ;Use FID at TG_7 with diameter of 0.000 feet.
        FID     1,      TG_8               ;Use FID at TG_8 with diameter of 0.000 feet.
        LOOK    FarFwd, FarSide,FarSide    ;Look using defaults favoring both sides.
        PANREL  Quick,  255, Ahead         ;Pan at Quick speed to Ahead relative azimuth.

        RUNON   Slow,   B13_A1
```

Figure 2.2. A partial SGV source program for a security robot

(Courtesy of Cybermotion, Inc.)

The beginning of an actual program for a security robot is shown in Figure 2.2. Notice that this program uses the concept of *including* files. In this case, the included files tell the compiler the numeric value for labels such as "Far", "Ahead", and "B13_A1". When compiled, this fragment would generate only 48 bytes of pseudo-code, which would be sufficient for the robot to reliably navigate the path segment. A segment could be as short as a few feet to as long as 256 feet.

Although programs such as the one in Figure 2.2 were originally written by a text editor, later versions were generated automatically by a GUI (Graphical User Interface). Paths are simply drawn on a map, and objects are dropped onto the map to define things the robot must know or do. The GUI then generates programs that can be edited by a standard text editor. Notice that this programming environment therefore borrows concepts from conventional languages, object-oriented languages, and visual programming.

Since a single instruction may elicit a complex behavior, scripting is ideal as a programming technique for robots.

The GUI revolution

As conventional languages became more complex, and development cycles became ever shorter, programmers began to covet more powerful ways of programming than typing into a text editor. At first, text editors were modified with special features that made them more efficient. For example, programming editors could call the compilers and linkers to speed testing of new code. The programmer was only made aware of mistakes when this was done, so many simple typing errors would still accumulate before being discovered.

Additionally, the early programmer was forced to either remember an incredible number of variable names and symbols, or to continually leave the editor to search for the desired label. This process led to many errors that would be discovered at compile time, and many that survived to plague actual operation.

> **Flashback...**
>
> I well remember an inspection I made in the early years of a new installation that contained thousands of handwritten robot programs. The programmers had been at the project so long that they had developed a "thousand yard stare" similar to that seen in soldiers who have been under artillery fire for a few months.
>
> As I stood watching a robot approach me, I became suspicious of its boldness in the confines of the area. At the last minute I realized it was singularly unimpressed with my presence and intended to run me over.
>
> Upon reviewing the handwritten programs we found that the programmer had substituted the label "fast" where the label "NC" should have been. As a result he had instructed the robot to execute the next 250 path segments without worrying about collision avoidance! At that moment I had an epiphany about GUIs, interlocks, and expert programming aids[2].

At first GUIs were developed for applications intended to be used by nontechnical people. As GUIs began to appear for these applications, the potential usefulness of the concept to programming environments quickly became apparent. Languages such as Visual C++ and Visual Basic appeared and gained favor. As an additional advantage, GUI interfaces had a metaphoric fit with the *object-oriented* language concepts that were to become dominant in high-level languages. The graphical interface was the perfect way to represent and manipulate objects.

---

[2] To quote Erwin Rommel, "Mortal danger is an effective antidote to fixed ideas."

The great rift

Interestingly, as main line conventional programming was faced with ever more complex requirements, it divided into two major philosophies under which many individual programming languages were born. The two major philosophies are conventional languages (such as C, LISP, and Basic), and database languages such as DB, and later Oracle and SQL.

Database languages were forced to confront vast quantities of data and primarily to search, sort, and cross reference data within these oceans of information. Conventional programs, on the other hand, evolved to facilitate applications ranging from word processing to video games.

As complexities continued to grow, these languages further evolved. Databases moved from using flat files to relational structures while conventional programs moved toward object-oriented structures. The C language became C++ and later Visual C++, while Basic adapted many of the best concepts of structured languages such as C, a GUI interface, and the concept of *objects*, and became Visual Basic.

Object-oriented programming

The highest of the fiefdoms we are going to consider is that of object-oriented programming. Such is the jargon surrounding this elegant and simple concept that its high priests have been able to discourage many mortals from even trying to understand it at a lay level.

The fact is, if you had never heard of object-oriented programming and simply spent a dozen or so years working on ever more complex conventional programs, you would either suffer a nervous breakdown or gradually invent most of the principles of object-oriented programming. Like so many of the concepts we will be discussing, this one is a natural solution to the problem of complexity.

Object-oriented programming is not so much a different way of programming as a different way of organizing your programs. It is an elegant structure that helps the programmer maintain order in the face of increasing complexity. Like all such concepts, it is more appropriate to some kinds of programs and less to others. It is very useful in complex programs such as those found at higher levels of control in mobile robotic systems.

Objects

So how is an object-oriented language different from a conventional language? Object-oriented languages still have the three basic elements of conventional languages, but they also have *objects*. Simply put, objects are groups of related functions and their associated local variables. With this metamorphosis, calls to the functions in an object are called *methods*, while local variables of the incorporated functions have become *properties*. An object is thus a useful metaphor for a chunk of code, as well as a way of building a protective shell around it.

Another principle of objects is that the programmer does not so much program an object, but rather provides a blueprint of how the object will be created. This blueprint contains code, as well as default settings for the properties. At run time, the program itself can generate objects using these blueprints. Thus, hundreds of instances of an object can be produced as needed by the application. Creating an object from its blueprint is called *instancing* it. When I opened this chapter in Word, I instanced a *document object* which you are reading now.

Programs are no longer static blocks of code. The actual executable code used by an object, along with all its variable storage and other resources is created and destroyed dynamically by the application. When an object is destroyed, the system must reclaim its memory allocation, or eventually all available memory will be used up. If this process is not completely efficient, the system will be said to experience "memory leaks," and applications may eventually crash as a result. Some versions of C++ were well known for this problem.

Properties

Properties are the local variables that control the behavior of an object, or that the object produces or acts upon. For example, if the object is a block of text, the properties might be *font*, *size*, *style*, and even *color*. This example demonstrates how appropriate the term *properties* really is!

The properties of an object cannot be tampered with from outside the object, but only by calling a method to set them. In this way, an object protects itself from having its controlling parameters set to illegal or dangerous values. Take the example of a program trying to set the *font* property of a text object. The desired font may not be available in the computer, so the object can substitute a default font and return an appropriate status code to let the caller know about the substitution. The result might not be what the programmer wanted, but it is better than filling the screen with smiley faces and arrows, or worse yet displaying the dreaded blue screen of death.

Methods

Methods are nothing more than functions within an object that act on the *properties* or perform tasks as modified by the *properties*. A call to a *method* almost always requires arguments to determine the details of its execution. In many cases, the arguments are not the value of variables, but rather the addresses of the variables. In this case, the argument is said to be "by reference" rather than "by value." The *method* may manipulate the values of these arguments by reference, or generate output in a myriad of other ways. In almost all cases, a *method* will return a status value that lets the calling program know if the desired action was successful.

Just as arrays of objects are possible, objects may contain other objects. In one example I remember, objects contained data types which in turn contained other objects, which in turn possess arrays.

The power of objects is that they allow us to build walls around well-tested blocks of code, and to use this code over and over with the certainty that it will execute safely. As an additional advantage, once an object has been created it makes envisioning the functioning of the whole module easier.

Robots and robot subsystems as objects

The important thing is to know when to use objects to represent something, and when to use more conventional representations such as arrays. A true zealot of the cause will opt for making everything in sight into an object. The result may be more confusing than helpful.

Generally, the more complex a block of code becomes, and the more variables and types of variables that act upon it, the better candidate it is to become an object. If you are writing a robot control program to execute on a PC, then you should give careful consideration to those items you wish to represent in this way. If you are expecting to allow third parties to access this code, then placing it into an object such as an OCX will be very useful.

Additionally, if you are going to include vendor-supplied code into your control system to support third-party systems like video compressors, the interface code will most likely be supplied to you as an object (e.g., OCX) file. In most cases this will allow you to integrate these systems very quickly, but if the object is poorly written there is nothing you can do to correct the problem.

Figure 2.3 is a list of a very few of the properties of an actual robot object as represented in its base-station control program. An array of these objects can be *instanced* when the program is installed (or upgraded), to support any number of robots. In this example, these properties are public to the other elements of the control program, but are not exposed to the outside.

If you haven't programmed in an object-oriented language before, it will take some time to begin to feel comfortable with how best to apply the concept. In Visual Basic, the programmer is given a great deal of latitude in the use of objects, and can make use of global variables to simplify communications with objects.

```
'Robot System Defined Properties:
Public Referenced As Boolean
Public Patrolling As Integer          'Off, Random, Sequential

Public PatrolMode As Integer
Public Recall As Boolean
Public CurrentJob As String
Public DestinationNode As String      'Destination of current job.
Public OriginNode As String           'Original starting node
Public CurrentNode As String          'Last destination node past
Public LastInsSent As Integer         'Last instruction in current job.

Public Comm As String
Public FilePath As String             'Dispatching Filepath
Public EventPath As String            'Event Logging File Path
Public SensorPath As String           'Sensor Logging File Path
Public TagPath As String              'Tag/Inventory File Path
Public RecordPath As String           'Recording file path
```

Figure 2.3. Some of the properties of a security robot
(Courtesy of Cybermotion, Inc.)

Network languages

Microsoft's most recent programming environments, VB Net and C' are intended to enhance object-oriented programming for networked applications. Under this regime, programs (or pieces of them) no longer necessarily execute in the computer in which they are installed, but rather programs may send objects to clients or other servers where they can be executed more efficiently.

Since the program may no longer all run in a single place, the convenience (and risk) of global variables no longer exists. There is likely a place for this concept in robot monitoring and control programs that link to networks, but its discussion is beyond the scope of this book.

The Basics of Real-time Software (For Mere Mortals)

Robots are real-time systems by nature, and to attempt to program one without an appreciation of the concepts of real-time systems is to play Sorcerer's Apprentice. Again, a mystique has developed around this field that is totally unjustified. The techniques used in real-time software are just extensions of the stack games played with interrupts. The beauty of a true real-time system is that it allows multiple programs (threads) to be written that look just like conventional programs. Again, we can best understand these concepts if we place them into an historical perspective.

Threads

Early PCs were fundamentally single-task machines. Under the CPM and DOS operating systems there was no concept of windows, so whatever program was running got all of the *resources* of the machine. Such an environment is referred to as being *single threaded*. When Windows 3.0 was introduced, it was not truly an operating system, but rather a DOS application that mimicked an operating system. Even modern versions of Windows like XP are not full-blown real-time operating systems, but they have enough of the attributes of such a system to allow them to mimic a real-time system.

Early PC operating systems forced the user to finish one task completely before loading the next. The result was very low productivity. Even so, the first elements of multitasking were already present in these systems in the form of interrupts.

Interrupts and stacks

Interrupts are a hardware mechanism designed to let the computer respond efficiently and promptly to asynchronous external events. The most common sources of interrupts are events such as a byte being received by a serial port, a disk drive finishing a sector write, a timer timing out, or some other event being completed.

An interrupt is like a computer's doorbell. If we did not have a doorbell in our house, we might find ourselves standing around the door for hours whenever we were expecting company. In a computer this is done by *looping,* and if it is done without releasing the processor it will absorb every cycle available. Essentially the program is asking "Is it here yet, is it here yet, is it here yet?"

Looping in a program without allowing other tasks to run is a crime against nature! Even if you expect the exit event to occur quickly, to put the program into a loop is to squander resources. If the exit event does not occur, the computer will appear to be locked-up and will not respond to the keyboard or other input.

Most modern processors have between 2 and 32 interrupt inputs. Most interrupt systems allow for expandability to larger numbers of inputs if needed. These inputs can be *masked* (disabled) under software control, and they can be either edge or level sensitive.

When an interrupt input is triggered, current program execution is stopped, and the address at which the program was executing is automatically pushed onto the *stack.* The *stack* is simply a block of reserved memory that is combined with a *Stack Pointer* to produce a simple LIFO (last-in first-out) buffer (see Figure 3.1). An interrupt is a hardware version of a subroutine call in software. It is just that simple.

When the return address has been pushed onto the stack, execution is restarted at the interrupt's preprogrammed *interrupt vector address.* When the interrupt routine is finished, it executes an *RTI (Return from Interrupt)*. An RTI is like a *Return* (from subroutine), in that it pops the return address from the stack and resumes execution of the interrupted program. The RTI usually has the additional effect of resetting any interrupt latch.

In practice, the interrupt routine will save all of the CPU registers on the stack as well as the return address saved by the hardware reaction to the interrupt. This is necessary because the interrupt routine may need to use the registers for its own calculations, and the interrupted program will need the data it was working on when its execution resumes. This is no more complicated than putting all your work onto a stack on your desk when a visitor arrives, and when they leave, taking it back off and resuming what you were doing.

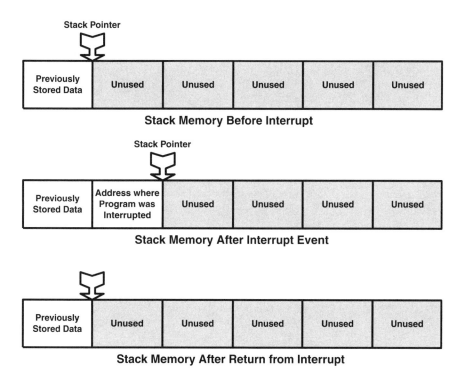

Figure 3.1. A simplified representation of stack operations during interrupts

The *interrupt vector* is the address to which the program will jump automatically when an interrupt occurs. This *interrupt vector* must have been set to point to a *valid interrupt* handler routine before the interrupt was enabled or the computer may begin trying to read and execute memory that is not valid. This has an effect similar to trying to assemble Christmas toys by discarding the instructions and reading the Kama Sutra instead; it is amusing but fraught with hazard. The technical term for this condition is *going off into the woodwork*. It can cause some very strange behavior.

I vividly remember an occasion on which there was a loud disturbance coming from a cubicle in the test area of the production floor. As I approached I saw the sensor head of an SR-2 Security robot moving erratically back and forth above the wall of the cubicle.

I recognized the curses coming from the cubicle as those of the company artist who was in a greatly agitated state. As he attempted to escape the small space, the robot repeatedly blocked his exit and threatened him like rabid animal.

After we subdued the robot by pushing one of its emergency stop buttons, we determined the cause of the event. Wanting to take some photos, the artist had turned on a robot that was undergoing a software upgrade and did not yet have a matching set of program chips installed. When the interrupts occurred, the robot's program "went off into the woodwork" and the robot tried to do the same!

This event caused us to build in a hardware latch that would detect illegal sequences and lock the computer. Most modern processors have special interrupts that occur if illegal instructions are encountered, but these won't do any good if they are not properly programmed.

Similarly, after years of entertaining Windows users with the "blue screen of death" and random acts of data vandalism, Microsoft finally incorporated good error trapping in its XP operating system. We like to think Bill was copying us.

Context

Interrupt routines can be state driven or they can maintain context. State-driven interrupt routines keep state data, like arrays and their pointers, to remember what they were doing, but the routine always starts at the top and ends at the bottom. If the program that uses the interrupt is to look like a normal program, and have exits when it needs to wait for something, it may switch stacks during the interrupt to do this. This is called *context switching*, and an interrupt routine that does this begins to look like a real-time kernel. The two programs on either side of this slight of hand appear to be completely independent of each other! I call this "going through the looking glass."

Consider the simple message decoder shown in Figure 3.2. This decoder expects a serial message that always starts with a "#" character, followed by a low and high character. The left side of Figure 3.2 shows how this might be accomplished with a state-driven interrupt. Note that even with this oversimplified example, the program flow under the state model already shows signs of having a lot of branching. Experienced programmers have an inherent abhorrence of this type of code, because they know it leads to chaos.

Now consider the program flow on the right side of Figure 3.2. The elegance of this program compared to the state-driven model is striking, and it will become more dramatic as the message protocol becomes more complex. The trick in the context model is that what appear to be simple *Get Character* calls are actually points where task switching occurs. Shortly you will see how simple this can be.

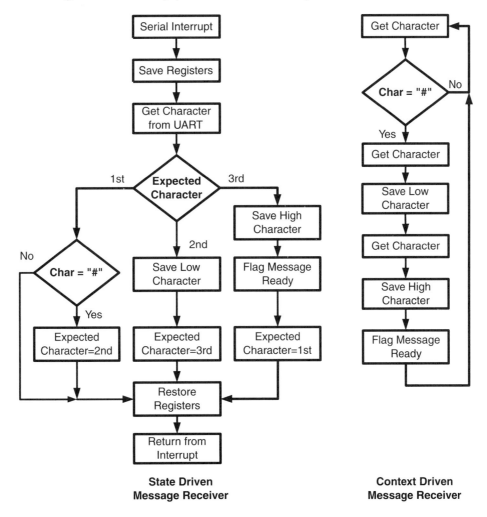

Figure 3.2. State-driven vs. context driven

Kernels and tasks

Real-time software concepts are almost as old as conventional programming concepts. A software program that allows the machine resources to be divided between multiple tasks is called a *real-time kernel*. The name *kernel* is appropriate because a kernel is the small but critical center of the programming environment. It is basically a task switch.

A *real-time operating system* contains a *kernel* and the basic disk, keyboard, mouse, and communications facilities found in conventional operating systems. Special microcomputer systems designed for factory automation sported simple *real-time operating systems* as early as the 1970s.

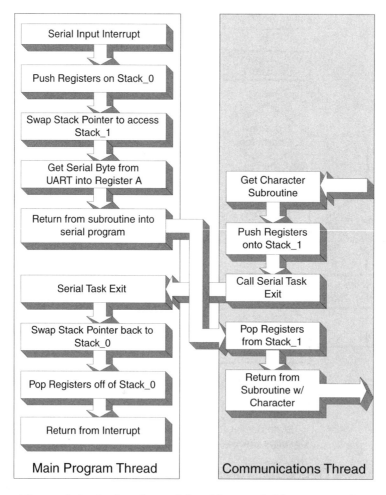

Figure 3.3. A simple multitasking serial input routine

The various operations that the computer has been programmed to perform are called *tasks or threads*. The kernel is the first code to execute when the computer is booted, and it essentially decides which tasks should receive the resources of the CPU and at what time. The *kernel* controls most of the interrupts of the system.

In most real-time systems, the kernel switches the CPU between tasks so rapidly that all of the tasks appear to be operating simultaneously. This is of course an illusion since the CPU can only execute one instruction at a time, but with CPU clock speeds in the gigahertz range, the CPU is virtually performing multiple tasks at the same time.

When the kernel switches tasks, it must do more than simply save the return address of the program it is interrupting. It must save the *context*. The context is the temporary data that the computer was manipulating to accomplish its task. This is most commonly the values of the CPU's registers, and any data already on the stack.

```
        ;------------------------------------------
        ;      Serial Input Interrupt Procedure
        ;
SER0IN:PUSH HL
       PUSH DE
       PUSH BC
       PUSH IX
       LD (SPMAIN0),SP ;Save main stack pointer.
       LD SP,(SP0)
       ;
       IN0 A,(RDR0)    ;Get received character.
       AND 07FH        ;Strip bit 8.
       RET             ;Return back into LOADER.
       ;
       ;------------------------------------------
       ;GETCHR is called from the serial loader
       ;program as if it were a subroutine used
       ;to wait for the next character. The
       ;the stack is switched back to the main
       ;program stack and the task is suspended
       ;until the next character arrives.
       ;
GETCHR0:PUSH IX
       PUSH HL
       PUSH DE
       ;  Back to foreground.
       CALL SR0EXIT
       ;  Return from foreground.
       POP DE
       POP HL
       POP IX
       RET
       ;------------------------------------------
       ;      EXIT from Interrupt level 1.
       ;      (Return to Foreground)
```

```
SR0EXIT:LD (SP0),SP
        LD SP,(SPMAIN0)
        POP IX
        POP BC  ;Restore registers
        POP DE
        POP HL
        RETI    ;Restore interrupts & exit.
        ;
        ;-----------------------------------------
```

Figure 3.4. Simple multitasking serial input routine coded for the Z-180 family

Figure 3.3 demonstrates just how simple a context switching can really be. The "looking glass" is the middle of the diagram. To the main program thread, the sequence on the left side of the figure looks like an ordinary serial interrupt. To the communications program thread, the sequence on the right side of the diagram looks like an ordinary subroutine. The communications thread will look like any ordinary program which calls the *Get Character* routine whenever it needs input. This is the magic routine referenced in Figure 3.2! In addition to making programs easy to code, not a single cycle of the CPU is wasted waiting for the serial character to come in!

The example of Figure 3.3 is shown coded in humble Z-80/Z-180 code in Figure 3.4. Unlike a conventional serial interrupt routine, the communications thread that calls *Get Character* maintains its *context*. That is to say that it can ask for a character until it sees the beginning of an incoming message, and then repeatedly ask for characters it expects to see if the message is valid as shown in Figure 3.2. If a complete message is received, or if invalid characters are received, it will loop back to looking for the beginning of the next incoming message. It looks like an ordinary single threaded program, but it only runs when new data arrives and releases the processor when it has processed that data and needs more.

Now if you have digested this bit of trickery, consider that the left half of the diagram does not necessarily represent an interrupt that occurs when the main thread is executing. The fact is it doesn't matter whether the main thread is executing, or another interrupt, or another independent thread. Context switching is entirely stack-based and therefore indifferent to sequence. Note, however, that the stack-pointers must be kept in dedicated memory storage that is reserved for this single event so that they may be restored during a context switch.

Note also that the stack pointers for all threads must be initialized to point at the top of their respective stacks before interrupts are enabled. Also note the initialization program must push the starting address for each thread onto its respective stack so that it will be available for the first "return" into the thread. Only the main thread, which starts at reset, does not require its stack to be initialized in this way.

Finally, notice that even though this process is stack oriented, such interrupts are *not* reentrant. For this reason, the interrupt that causes the context switch must remain masked (disabled) until the context has been restored. In most systems, this is the default and the programmer would have to reenable the interrupt intentionally to cause a problem.

If you are going to write programs for dedicated microprocessors that do not have operating systems, you need to master this technique. If you are using a system that has context switching, it is important to know how it works in order to use it most efficiently.

Task switching

A real-time kernel can be extremely simple or quite complex and if you are working with small microcomputers you can readily write your own to fit your needs. There are many ways in which the kernel can interrupt and resume execution of various tasks. The most obvious way is in response to interrupt events.

Interrupt events

The serial communications task interrupt described above is an *event*. Events can cause the kernel to divert control to a specific task. The task must then relinquish the CPU by calling for another event (going back through the looking glass). Timers can also create events. In more complex kernels the task can relinquish the CPU while asking for one or more events to restart it.

For example, a communications task might ask for a character from the serial port OR a timeout of 1 second, whichever comes first. If the timer expired first, the interrupt would return a special flag to the calling thread to indicate that the event was a time out and not a valid character being received. The beauty of event-driven task switching is that it is very efficient. Only threads with work to do get the CPU's resources.

The kernel thus stands between the interrupts and the threads. Program flow thus goes from the interrupt to the kernel, and then through a context switch to a consuming thread if and when it is expecting the data. If your system is very simple and resources like serial ports are not shared between threads, then a true kernel may be unnecessary.

Time slicing

Time slicing is a simple method of sharing CPU resources between tasks. Unlike the above interrupt event example, the tasks are not required to relinquish control, it simply happens to them at any time in the execution of their program. Threads may be allowed to mask the switch during critical operations, but otherwise the sharing operation is totally asynchronous to their operations.

Basically a time slicing kernel receives an interrupt from a hardware timer, which interrupts the thread that is currently running. The registers for that thread are saved on its stack, and the kernel switches the stack to the next task. Execution continues until the next timer interrupt, when the process is repeated.

UNIX and other quasi real-time operating systems have time-slicing task switching. In most of these systems the time slice can be set larger for some tasks and shorter for others, giving important tasks a bigger slice of the CPU pie.

Occasionally there will be multiple threads that do not lend themselves to *event* task switching, while other threads require time slicing. In these cases, time slicing can be used on some tasks while event switching can be used on others. The biggest shortcoming of time slicing is that it does not necessarily provide optimal allocation of CPU resources. Additionally, timing glitches may be induced in the tasks being switched because time slicing is not generally synchronous with any particular place in these threads. So if another process is consuming their output, the programmer must assure that that process does not get half finished results.

Normally, the kernel manages the list of running threads, the priority for each thread and the status of the thread, the events it is waiting for, and so forth. For small, dedicated systems, all of this overhead may not be justified, and the programmer can just write a simple time slicing context switch driven by a timer interrupt. The trade-off is one of efficiency versus flexibility.

Reentrance

In a real-time system, the programmer no longer defines the exact sequence in which code will execute. Instead, the programmer sets the ground rules, and events determine the way things go down. This fact requires the programmer to go from linear thinking to thinking in dynamic terms.

The first difference becomes apparent when programming functions (subroutines) that are general purpose in nature. These routines may be called by many different threads. The problem comes when one thread is using a subroutine and the kernel switches the context to another thread which then calls the same routine. Since the function had not completed its work for the first thread before being called by the second, it is said to have been *reentered*. A function or subroutine that can allow this without errors is said to be *reentrant*.

If a routine uses any dedicated variables in fixed memory locations, the value of these will be destroyed by a reentrant call, and when the routine returns to finish the first call it will produce erroneous results. Therefore, for a routine to be reentrant, it must use only registers which are preserved on its stack, or stack allocated memory for its working storage.

When at all possible, all functions and subroutines should be written to be reentrant. Reentrant routines should also be clearly labeled as being so, and conversely non-reentrant routines should be clearly marked as well.

If a subroutine is not reentrant, and it is subject to being called from two or more different threads or interrupt routines, then the only solution may be to mask interrupts during the subroutine. Generally this is a messy solution that can bring with it other problems. For example, if a very large block of code is being executed with interrupts blocked while fast interrupts are occurring, there is the danger that one or more interrupts may not get serviced. This condition is called *overrunning* the interrupts.

Interrupt masking and interrupt priority

There are other considerations when programming real-time systems. For example, consider an interrupt from a wheel encoder that is used to make dead-reckoning calculations. At least two things can go wrong. The first danger has to do with *interrupt priority*.

The encoder procedure will be accumulating position estimates based on previous position estimates. For this reason it cannot be reentrant by definition, as it needs the results of all previous executions to be complete before it can add the new increments to their totals. Therefore this routine must mask its own interrupt (or simply not reenable it) until it is done with its calculations.

Such an encoder will produce interrupts that can occur at a relatively high rate, but that take only a few microseconds to service. Since the shortest time period between encoder interrupts is much greater than the length of time required for the interrupt procedure to execute, one might think there could be no risk of an interrupt *overrun*. Wrong!

What if this encoder service routine itself gets interrupted by an interrupt with a longer service cycle? If this is allowed to happen, then the encoder service routine may not complete its calculations before another interrupt comes in from the encoder. Since the interrupt is masked, it will miss this encoder tick. Such an error may be so small that it is never noticed overtly, but the dead-reckoning of the robot will be significantly degraded.

For this reason, all interrupt controllers offer a way of setting *interrupt priority*. An interrupt of a higher priority can interrupt an interrupt handler (or its thread) of lower priority, but the lower priority interrupt will be automatically *held off* while any higher priority interrupt is being serviced. In such a case the lower priority interrupt will set a latch, so that it will be serviced when all higher priority interrupts have been serviced.

Many errors that occur in real-time systems are subtle in their symptoms. The best protection against them is the forethought of the programmer.

Inter-task communications

The second potential problem with our dead-reckoning interrupt procedure originates from the way we save our results. For example, the dead-reckoning routine will be calculating both X and Y coordinates and possibly Azimuth. If it saves one of these and is interrupted before it can save the others, then an interrupting thread that consumes this data may pick up a garbled set of coordinates. Similarly, if a consumer of these coordinates has read the X value and is about to read the Y value when the encoder interrupt occurs, it may end up with a mismatched set.

This sort of problem can be even more severe if a routine must store more than one byte or word, all of which represent a single value. Partial storage in such cases can cause very large errors because of the nature of binary data storage. To avoid these problems, the programmer may get the entire set of results prepared in local storage, then *hold off* all interrupts for the few cycles that it takes to store the complete set.

Here the power of the object-oriented concepts of the previous chapter play into the concepts of multitasking. If we build a *dead-reckoning object*, then we can force other tasks to call its methods to get things like the latest position estimate and we can build the necessary protection into the object itself rather than exposing public variables.

Visual Basic and real-time controls

It is obviously beyond the scope of this book to attempt to address the basics of Visual Basic programming. There are many fine reference books on the subject, but they do not generally talk about real-time considerations, so I will try to provide some useful insights here. Although Visual Basic (VB) is not a true real-time language, it is a remarkably easy-to-use object-oriented language that can do a very impressive job of faking real-time operations. To understand how this can be done, consider how VB and Windows switch tasks.

VB events

VB depends upon *events* to allocate CPU time to different tasks. These events are somewhat different from those used by conventional kernels in that most are more software in nature as opposed to being interrupt driven. The operating system maintains a list of all the possible *events* that could occur at any given time. VB does have *OnComm Events* that support basic serial communications interrupts, as well as other events that are linked to hardware interrupts, but on the whole, most events are software generated.

For example, as a program posts a form on the screen with controls (buttons, sliders, etc.), the *events* that have been programmed for the controls are added to the list of possible events. If the mouse is then clicked, for example, the operating system will evaluate the cursor position to see if it is over an active control, and whether there is a procedure listed for clicking that control. If there is a procedure for the control, and if the control is enabled, then the procedure is called (executed).

DoEvents

The mouse itself may generate interrupts as it is moved, but under Windows, these do not elicit macro-events. The operating system instead simply stores away the position of the cursor. This cursor information only gets accessed and evaluated, and action taken, when the operating system has control of the CPU. But if the interrupt did not provide this control to the operating system through an interrupt, then how did it get control? The answer is that all well-behaved programs must release control to the operating system on a frequent basis.

Many functions are called infrequently, execute quickly, and release the processor. However, other threads may execute for substantial periods of time as they act on large amounts of data or loop waiting for a condition to occur. These tasks must release the processor (CPU) periodically. In VB this is done by including a function call to *DoEvents*.

DoEvents is an operating system call that relinquishes control of the CPU to the operating system so that it can determine if an action event has occurred. If an event has occurred, it will be serviced and then the CPU will return to continue servicing the original thread.

For example, the operating system might find that a hardware timer has timed out and thus execute a function associated with that timer. *Under VB, individual timers are not implemented with hardware counters as they are on many microcontrollers. Instead, they are more like the settings of an alarm clock against which a master clock is compared. Therefore, a timer does not time out and cause an action, but rather the operating system calculating that the timer has timed out causes the action event.* This important distinction has both an advantage and a disadvantage.

Since the operating system must gain control of the CPU in order to execute any event, threads are in no danger of being blindsided by an interrupt and losing control of the CPU involuntarily. Instead threads must complete and release the CPU, or they must intentionally relinquish control temporarily by executing a DoEvents function call. This important distinction makes writing well-behaved tasks much simpler under VB. This is akin to masking off interrupts, or more accurately only enabling interrupts at specific points.

Unfortunately, like training wheels on a bicycle, this protection comes at a price. The price is that no timer or other software event will be serviced instantly, but it will instead be recognized and executed the next time some polite running thread releases control of the CPU.

The longest period between when the event actually occurs and when it is serviced, is called its *latency*. If threads are written with very large blocks of code that are not broken up by DoEvents calls, then the latency can be as long as the time required for the largest of these blocks to execute. This means that timer-driven tasks under VB are not very precise, especially if other threads are doing time-consuming tasks such as graphics. Even so, with CPU speeds of several GHz, the latency may not be very severe.

Also, it is important to notice that VB does not offer any ready way to prioritize events. If a call to DoEvents is made by a thread, the next active event on the list will be serviced first, regardless of its importance. This fact adds to the indeterminate nature of event latency under VB. As a benchmark, I have written elaborate 8-channel PID algorithms that occurred on 16ms intervals in an environment supporting modest graphical animation, and experienced manageable latency.

So when should a DoEvents call be placed in a thread? The answer comes in understanding the kind of things that bog down a CPU. In VB, these include loops, handling of large arrays or strings, graphical operations, and screen-related functions such as changing the background color of a text window. In each of these cases, the function should call DoEvents immediately before and after such events, or in the case of loops it should be done within the loop.

```
'Post a message form with yes, no, and cancel buttons.
'This form will return the operator's selection through
'the global variable Response.

    YesNoCancel xYellow, "Are you sure it is safe to operate?"

    Do While Unloading = False And Response = 0
        DoEvents
    Loop

    If Response <> YesResponse Then
        Exit Sub
    End If

'Now that we have permission, let's start the system.
```

Figure 3.5. Using DoEvents in a loop

Figure 3.5 demonstrates several very important considerations required when using VB as a real-time system. The first line calls the custom function *YesNoCancel*. The function YesNoCancel launches a form which presents the question text on a colored background and which has buttons for yes, no, and cancel. When the form is launched, the public variable *Response* is set to zero to indicate that a response has not occurred.

The program then enters a loop to wait for the operator to select one of these buttons. If a response occurs, the result is placed in the public variable *Response*, and the query form closes. As a matter of interest, the query form also has a timer which will enter a *Cancel* response and close the form if the operator does not respond in a reasonable time.

Notice that if the DoEvents call is not included in the loop, then the loop will not release the CPU, and the query form will not receive any CPU time. As a result the events of the operator clicking any of its response buttons will never fire, and the whole program will lock up waiting for something that will never happen.

In this example, if any response other than a *yes* response is returned, then the program terminates and takes no action. This is all quite clear, but what is the variable *Unloading* all about?

Freddy as a VB form

In the famous thriller *Friday the 13th* and its many sequels, the demented killer Freddy refuses to stay dead and continues to pop up to cause mayhem. VB can create forms and functions that use Freddy as their role model. The problem usually occurs in MDI (*Multi-Document Interface*) applications.

An MDI consists of a main form with a menu at the top, and it has any number of *daughter* forms that may appear within the main form's work area. Microsoft Word is an example of an MDI. The MDI format is ideal for real-time controls that have multiple forms.

The problem comes when a well-behaved MDI daughter form or function calls *DoEvents*, and the event that occurs is a closure of its *Main* form. Closing the Main form automatically closes any of its open daughter forms and functions, but this means that the very program thread that called *DoEvents* is gone when the operating system attempts to return into it after executing the closure! This situation will crash the computer, so the operating system resurrects the snippet of code or the form that called *DoEvents*. This results in either the form popping back up after the parent MDI is gone, or in the case of a function it may simply resurrect an invisible chunk of code and allow it to continue to run invisibly. In some cases, it may even resurrect the MDI *Main* form.

This bug is present in a lot of shareware programs, and not a few commercial applications. After you close the program, if you look at your system *resources*, you will see that a part of it is still running. Even if there is no visible element left open, the running code absorbs machine cycles (resources), and can cause problems when shutting down Windows.

Luckily, VB has enough flexibility to allow a work-around for this problem. Any form, including the *Main* form, can include a subroutine called *Unload*. If the form has such a subroutine, then it will be called if an attempt is made to close the form. The argument *Cancel* is passed to the *Unload* subroutine by reference, and setting it true will cancel the *unload event*.

One work around for the Freddy problem is to have the *Unload* subroutine use a sort of scuttling charge in the form of a timer. If the timer is not already running, then *Unload* will start the timer, set *cancel* true, and set a public flag called *Unloading* to true. The result of this sequence is to delay the *unload event*, and to warn all other threads that may be calling *DoEvents* to stop doing so and move on. When the timer goes off, the *Unload* routine will fire again, but this time it will not cancel the unload event. A timeout of a second or two is usually plenty.

```
Private Sub MDIForm_Unload(Cancel As Integer)

    If Timer1.Enabled = False Then

        Unloading = True          'Flag other tasks.
        Timer1.Enabled = True     'Set scuttling charge.
        Cancel = True
    End If

End Sub

Private Sub Timer1()              'Scuttling charge has fired.
    Unload me
End Sub
```

Figure 3.6. Delaying the main form unload

The simple *Unload* subroutine of Figure 3.6 is all that is required to accomplish this delayed shutdown sequence. Note also that if a form contains an unload function, the next instruction must be the end of the subroutine execution. The following example *will not unload properly* because there is executable code after the unload function.

```
Private Sub Framus(unloadflag as boolean)
    If unloadflag=true then
        Unload me
    End if
    <more code>
End sub
```

To correct this problem, simply add the line "exit sub" immediately after the "Unload me" and before the "End if."

Modal controls

If you have programmed in VB before, you probably had another question about the code in Figure 3.5—namely, why go to all the trouble to create a query form like that launched by *YesNoCancel?* After all, VB has a standard control called MsgBox whose style can be set to *YesNo* to perform exactly this function. Well, not quite…

Many VB controls that require input are programmed to intentionally block their thread (and thus all other events) until they get that input. In other words, they are not polite and do not call *DoEvents* while waiting for their input. They are said to be *modal* and cannot be used in a real-time application, because they will block all threads and background tasks. Exactly why they were not given a property that could be changed from *modal* to *non-modal* is a bit of a mystery, as many very nice controls are rendered useless to real-time programmers.

Some other tips on using VB for real-time applications

Many VB controls like text boxes are not quite as smart as one would expect. For example, if the text in the box is "System On," and a line of code again sets the text to "System On," the text will be repainted despite the fact that nothing has changed. The same is true if the background color is set to the same value that it already is. This will cause these controls to flicker and it will waste CPU cycles.

When a large number of readings, bar-graphs, and other visible controls need to be refreshed regularly, this can cause a very significant waste of computer resources. For this reason one should include code that tests to see if the new text or property is different from the old one before attempting to set it.

Setting up a structure

As we discussed in Chapter 1, the most important decisions you will make are those regarding the structure of your system. Here are some pointers to remember.

1. List all of the services your program will need to provide. These might include optional and transient services as well as full-time support services such as communications interfaces.

2. Decide which of these will go into common threads and rough in a structure.

3. Decide on the ways you will handle task switching.

4. If you will have control of interrupts, decide on the interrupt priority order that will assure critical services get done first. If programming in VB, make sure no task hogs CPU time.

5. Think about which functions are generating data, and which are consuming it, and look for timing conflicts.

6. Write critical components like communications drivers first.

7. Test each component and service as it is added.

8. Build in error handling at every level, and think about how you are going to monitor the system's performance.

9. Decide on the units you will use for distances and angles, and create a basic library of mathematical and geometric functions.

Creating a library

A good library of mathematical and geometric functions is an essential part of the foundation of a robot control program. This library should be written with speed and reentrance in mind. This means that library functions must keep all variables locally, and units of measure and conventions must be consistent between functions.

Decide the type of variables you will use, and be consistent. If angles are assumed to be positive for the clockwise direction in one routine, then all routines must assume this. If the system is to accept both English and metric units of measure, decide which system it will actually use, and make the conversions to the other system at the operator interface.

If you are programming in a high-level language, then the choices will range from integers and long integers, to singles and doubles. Remember that nothing is free. If you decide to use singles instead of integers, you may slow the calculations dramatically. On the other hand, if your target system has a math coprocessor, this might not be the case.

Mathematical functions

Generally you will need only the basic mathematic functions for the techniques described later. These include signed addition, subtraction, multiplication, division, squares, and square roots. Note that whatever variable types you choose, interim calculations may overrun the type limits. Routines must therefore be able to handle this internally without returning errors.

Geometric and vector functions

In the coming chapters we will discuss various processes in terms of adding vectors to each other and of converting vectors to Cartesian coordinates and vice-versa. For clarity and brevity, the mathematical details of how this is done will not be included in the discussions. The library must support this capability. For many indoor applications, a two-dimensional frame of reference is completely adequate, but for outdoor systems the library must include the Z axis in all vector functions. At a minimum, the library must allow support for the following very basic functions:

- Add one or more vectors together and return a single vector.

- Add one or more vectors to a coordinate and return the resulting coordinate.

- Calculate the vector between two coordinates.

- Calculate the included angle between vectors with the same base.

- Calculate the angular sum between vectors with the same base.

If integers are to be used, then all functions should take rounding into consideration to achieve maximum accuracy. If this is not done, a phenomenon known as *digital integration* can occur, which can cause significant error in long-term results.

Once a basic library is produced, it can be supplemented as required. Before I code anything in a routine, I find it useful to include a paragraph or so of comments defining what the routine is to do, what variables it uses, what units they are in, and anything else that a user might need to know about it. These comments are useful both in structuring the routine and in later maintaining and using it.

Create a configuration definition

If you want your program to be maintainable, then *do not code constants into any part of it*! This means that parameters such as distance between the wheels of the robot, or the position and orientation of each sensor should be variables that are preset in operation. When something in the configuration changes, it will be a simple matter to correct all the calculations that use this measurement by simply changing the single variable.

Never code anything specific to a configuration into your library.

Test your library at every boundary

One of the best places for software bugs to hide is near boundaries. Test all your routines with maximum and minimum input values in all combinations of signs and magnitudes. It is even useful to write a simple automatic test program which increments the inputs linearly and graphs the output of each function. *Mathematical glitches in these functions will be very difficult to diagnose once you start your system running.*

CHAPTER 4

Thinking More Clearly Through Fuzzy Logic

Of all the software concepts touched upon in this book, fuzzy logic is one of the most valuable and most widely misunderstood. These misunderstandings are at least in part a consequence of the fact that the term *fuzzy* is not transparently descriptive. The basic concept of fuzzy logic is so simple that it amounts to little more than common sense. If you understand the mathematics (geometry) behind a straight line, you can master fuzzy logic.

The purpose of fuzzy logic is to provide the best possible "guess" at the value of something that cannot be measured directly, but which can be inferred from a combination of inputs. For this reason, a system that accomplishes this is sometimes referred to as an "inference engine."

To understand the place of *fuzzy logic*, we will first consider Boolean logic as it has been used in sensor based systems. In Boolean logic, all inputs, outputs, and calculations deal with two states; 'true' and 'false'. These states may be expressed as '1' and '0', or –1 (FFFFh) and '0', but they always represent the two Boolean states. If inputs are not Boolean to begin with, then they are compared to a *threshold* to convert them to Boolean states before being manipulated by the Boolean logic.

A simple example of this process can be seen in a "dual technology" motion detector used in the security industry. In the 1970's and early 1980's, several motion detector technologies vied for market supremacy. These included light-beam, ultrasonic, passive infrared (PIR), and microwave based systems. The benchmark for effectiveness was measured by how sensitive the system could be set without experiencing an unacceptable rate of false alarms. The PIR and microwave technologies eventually emerged as dominant, but both could be false alarmed under some circumstances.

A microwave motion detector emits a UHF radio signal which reflects off objects in the environment and is then received back at the detector where it is mixed with the outgoing signal. If nothing moves in the environment, the two frequencies are the same and no sum or difference "beat" signal is produced. However, if some of the returning signal was reflected from a moving object, then a frequency (beat note) is produced that is proportional in frequency to the velocity of the target. This signal is then amplified, rectified, and compared to a threshold. If it is greater in amplitude than an adjustable *threshold*, then a relay closes to signal an alarm.

A PIR alarm, on the other hand, is purely passive. Black body radiation (including body heat) in the 8 to 12 micron wave lengths is passed through a Freznel lens that has multiple facets. The images from these lenses are focused onto two elements of a detector that are connected in opposite polarity. As the heat source moves, the dancing images on the detector cause a pulsating signal. This signal is then amplified, rectified, and compared to a threshold just as is done in microwave systems.

Microwave systems can be falsely triggered by stray signals from other systems, and such systems could actually detect large moving vehicles outside of the building that did not represent true intruders. Passive infrared sensors can be falsely triggered by heat from forced air heaters, as well as by blowing curtains and certain other signal sources. Initially, manufacturers attempted to reduce false alarms with signal-processing techniques such as time delays. Despite these marginal improvements, the market wanted a system whose false alarms were much less frequent than either system could achieve.

As the market grew, eventually the cost of these systems became low enough that it was possible to incorporate both technologies into a single alarm without the price becoming prohibitive. The result is shown in Figure 4.1. Note that the symbol shown as an "Amp" is both an amplifier and detector, which puts out a fluctuating DC signal proportional to the disturbance magnitude.

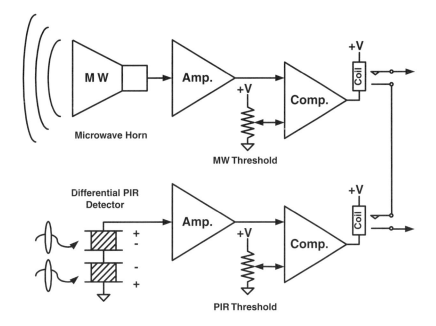

Figure 4.1. A dual technology motion detector

Since the disturbances that cause PIR technology to false alarm are not related to those that cause microwave technology to false alarm, the false alarm rate of a combined system is much lower than for either technology alone. The Boolean equation for an alarm is stated as:

Alarm = (MW > MW Threshold) AND (PIR > PIR Threshold)

Or:

Alarm = (MW > MW Threshold) · (PIR > PIR Threshold)

Notice that the comparators convert the analog signals to binary (on/off) signals. Interestingly, most early systems actually contained two relays. The Boolean AND function was implemented by wiring the contacts of these relays in series. Why this was not done with simple logic is puzzling, as the extra relay is much more expensive than a couple of transistors or even diodes. Even so, the dual technology alarm was an immediate success and has remained a mainstay of the industry.

The dual technology detector is a classical example of Boolean logic in a signal processing application, but it had a logical weakness. To understand this weakness, let's arbitrarily call a signal 100% when it is precisely at the threshold level. Now consider a case when the microwave signal is 99% and the PIR signal is 200%. This would not generate an alarm, yet it would be more likely to represent a true detection than the case when an alarm would be generated by both signals being 101%. There is a better way to combine these signals.

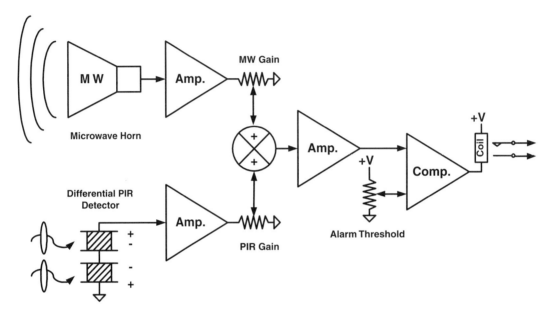

Figure 4.2. Dual technology motion detector using fuzzy logic equivalence

The circuit of Figure 4.2 demonstrates how we might sum the two sensor signals of the motion detector before comparing the result to a threshold. Instead of converting the signals to Boolean values before ANDing them, we sum their analog levels and then compare the result to a threshold. Two gain potentiometers have been provided so that we can balance the two signals to appropriately represent their relative sensitivities.

If the final threshold is set higher than either sensor signal can achieve by itself, then both signals must be present to at least some degree to cause an alarm. The advantage to this approach is that a much wider range of valid alarm input combinations is possible. *What we have done is to create an electrical equivalent to a fuzzy logic solution.*

Now if we were to read the voltage out of the summing amplifier with an A/D (analog to digital) converter instead of feeding it to the hardware comparator, then we could consider this signal to be proportional to the intrusion threat. We could compare this threat level to software thresholds and perform all alarm processing in software. Better yet, if we moved back to reading the outputs of the two signal amplifiers, and perform the gain balancing, summing, and thresholding in software, then we have created a true *fuzzy logic* intrusion detector. What could be simpler?

Trapezoidal fuzzy logic

The simplest form of fuzzy logic is a linear technique referred to as *trapezoidal fuzzy logic*. This technique allows a mere handful of variables to determine the contribution functions of each channel.

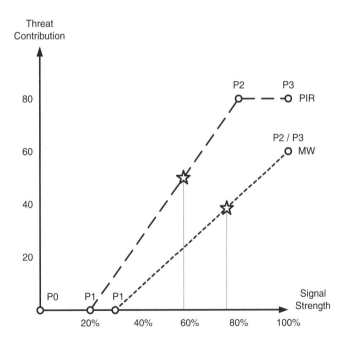

Figure 4.3. Trapezoidal fuzzy logic

Trapezoidal fuzzy logic is nothing more than a way of defining the contribution of inputs to an output using only straight line functions. Each point is either an end point, or it is the intersection of two adjoining line segments.

Let's assume that we have digitized the outputs of the detector / amplifiers of Figure 4.2 and scaled them as 0–100%. To combine these signals into a single "threat score," we have only to define a series of line segments that *monotonically* express the relationship as shown in Figure 4.3. By *monotonic*, we simply mean that for any value of signal strength there is one and only one value of threat contribution, or in other words the curves don't double back over themselves.

In this case, the PIR signal transfer function is defined by P0, P1, P2, and P3, whereas the MW signal transfer function has only P0, P1, and P2 as distinct points. For purposes of calculation, the third point (P3) can be simply defined as having the same value as P2. Notice that there is no absolute rule for the number of points, and if need be the functions can have many more line segments.

In the case of our motion detector, we have decided that we don't want the PIR to contribute to the threat score if it is indicating less than 20% of full scale, because signals below this range are commonly experienced as background noise. This is referred to as its *dead band*, and is the range between P0 and P1. The MW sensor is even noisier, so we have given it a *dead band* of 30%. The dead band assures that we cannot produce an alarm state with one sensor in a region so low that it only represents normal background noise.

The highest threat contribution we have allowed for PIR is 80 units, and the highest for MW is 60 units. Thus, if we set the alarm threshold to 75 units, the PIR alone could cause an alarm if it approached 100% of full-scale input, but the MW could not produce an alarm by itself. However, if we set the threshold above 80 units, neither sensor by itself could cause an alarm, and thus both would need to be above their P1 points.

Consider the case where the two sensors are responding at the levels indicated by the stars. The PIR is contributing about 58 units of threat, while the MW is contributing 75 units. Thus, the threat score is 58 + 75 or 133 units. If this is greater than our alarm threshold, we have an alarm condition.

Fuzzy democracy

Perhaps the most frustrating quality of democracy is that everyone gets one vote, whether they understand the issues at stake or not. A person with a great deal of insight into an issue can be cancelled out by a person who simply likes the charisma of the opposing candidate, or who votes a straight party line.

But what if we used voting machines loaded with a thousand questions about current issues, all of which were published ahead of time? Each voter could be presented with ten of these questions, randomly selected from the thousand. Upon finishing the quiz the voter would then vote that score to the candidates of his or her choice.

Such a concept is unlikely to be implemented in our society (largely because the politicians who make laws find charisma easier to feign than competence), but fuzzy logic has that capability, and much more. The earlier example of a dual technology motion sensor favored the passive infrared sensor because it is generally a better indicator of a true alarm. Yet because the two sensors contributed proportionally to their signal strength readings, the microwave could have more affect on the outcome than the PIR. Thus fuzzy logic is itself a sort of democracy. The process can, however, be taken to higher levels of processing.

In the coming chapters we will explore ways that the results of fuzzy calculations can themselves be inputs to other layers of fuzzy logic. In the end, all of these concepts are tools that the creative programmer can put together in endless combinations. The power of this layering of simple concepts is hard to overestimate.

Adaptive fuzzy logic

The biggest advantage of software controls over their analog counterparts is that the rules for signal processing can made be quite complex without adding to the system's cost. Since only a few parameters control the operation of a fuzzy logic "engine," it can be easily slaved to other controls to gain a degree of adaptivity.

Weighting trapezoids in response to other parameters

Returning to our example, as the ambient background temperature approaches the temperature of a human body; the PIR sensor will become less sensitive because there is no contrast between our intruder and the background. Any signal the PIR does produce may need to be weighted more heavily to prevent our alarm system from becoming insensitive. We could thus add a second axis to the fuzzy logic transfer function that looked like Figure 4.4.

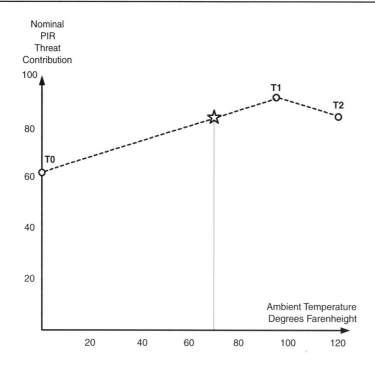

Figure 4.4. Weighting function of PIR over temperature

By defining the three points T0, T1, and T2, we have determined the sensitivity of our earlier profile to temperature. Notice that once the background temperature passes body temperature, our intruder will begin to stand out again because he is *cooler* than the background. For this reason, the "gain" is reduced above 96 degrees.

To perform the whole transformation, we simply read the temperature, use the graph of Figure 4.4 to determine the nominal threat contribution for the PIR sensor at the current temperature, and then set the contribution value of the points P1 and P2 in Figure 4.3 to this value.

The star in Figure 4.4 represents the nominal contribution for 70 degrees, the same operating point as used in Figure 4.3. This could all be shown as a three-axis graph, but since it will be processed as just described, it is less confusing to show the temperature sensitivity as a second graph.

On the other hand, an environment might be very quiet to the microwave sensor, indicating it could be trusted more heavily than the factory default setting would imply. In this case, we might "learn" the background noise level over a period of time, and then weight the MW peak threat contribution accordingly.

Multipass and fratricidal fuzzy logic

As was mentioned at the beginning of this chapter, the purpose of fuzzy logic is to infer the value of something that cannot be measured, from a combination of related things that can be measured. A passive infrared sensor does not measure intruders, it measures heat disturbances. A microwave sensor does not measure intruders either, but rather the presence of Doppler-shifted microwave reflections.

In the previous example, we have used fuzzy logic to *infer* the intrusion threat probability from these two sensors. We could easily add additional sensors at will. If a sensor did not detect the intruder, it would simply not contribute to the threat score.

There are some kinds of calculations, however, where this may not be preferable because we are trying to determine a specific value. In these cases, we do not want a sensor to throw an "outlier" into the calculation if it has nothing competent to add. One way to avoid this is to calculate the contribution for all sensors on a first pass.

Once this has been done, we can calculate the "average" inference of all the sensors, and then determine if any sensors are widely out from the consensus of the others. If this is the case, then we can throw out the odd-ball and recalculate without it. I used the word "average" in quotes, because in some cases it may be best to use a median, or RMS (root-mean-square) value as the consensus.

> *Flashback...*
>
> I am reminded of an example of just how powerful simple fuzzy logic can be. It was early evening and I was about to log into a warehouse installation of a security robot that we had completed a few weeks earlier. It was standard practice to download log files and see if anything needed tweaking
>
> Just as I was connecting, the operator called me on the telephone. He told me that the robot was false alarming on fire threat, that he had dispatched a security officer twice, and that there was no fire threat at all. Could I make it stop false alarming?
>
> When I logged into the console the most recent alarm had been cleared, but I saw that the explosive gas sensor was reading very high. Suspecting a defective sensor, I look at a graph of its recent history and was alarmed to see it fluctuating from almost nothing to almost full scale. Since this is not a failure mode of such sensors, I looked at the alarm report and saw that the robot had alarmed because of a high gas reading in the presence of an open flame.
>
> Next I downloaded the sensor history for the shift and plotted it on the warehouse map. The pattern showed a huge dome of gas. But what about the flame? Having digital video,

the robot had pointed its camera at the threat and recorded a video clip as part of its alarm report. When I downloaded the clip, I saw the culprit. There, on the very edge of the gas dome was an overhead gas heater with its pilot light burning. The console had actually announced the alarm stating that it was from gas with a strong flame contribution, but since the officer could not smell the gas, and didn't think of the heater as a flame, he had dismissed the threat.

Summary

There are many ways to control the weighting in a fuzzy logic inference engine. It is important to think of fuzzy logic as a mathematical process that can be embedded into larger structures, rather than a stand-alone discipline.

This humble mathematical process can provide very impressive decision making when it is properly employed in conjunction with other techniques. In the chapters to come, we will see how merging fuzzy logic with modeling and other concepts can produce spookily powerful results.

Closed Loop Controls, Rabbits and Hounds

Any good athlete will tell you that the key to an exceptional performance is to imagine the task ahead and then to practice until the body can bring this imagined sequence into reality. Similarly, generals create "battle plans," and business people create "business plans" against which they can measure their performance. In each case there is a master plan, and then separate plans for each component. An athlete has a script for each limb, the general for each command, and so forth. All of these plans must be kept in synchronization.

Controlling a robot or other complex intelligent machine requires that it have a plan or model of what it expects to accomplish. Like all such plans, it is quite likely to require modification within moments of the beginning of its execution, but it is still essential.

Like the commander or executive who must modify a plan to allow for changing situations, a robot must also be able to modify its plan smoothly, continuously, and on the fly. If circumstances cause one part of the plan to be modified, then the plans for all other parts must be capable of adjusting to this.

For the plan to have any meaning, physical hardware will need to track to the plan to the best of its ability. Thus we trespass into the fiefdom of *Controls*. This subject is deep and broad, and engineers spend their entire careers mastering it. In short, it is a great subject to trample on. The fact is that it is quite possible to create decent controls for most applications using little more than common sense algorithms.

The purpose of a *control system* is to compare the plan to reality, and to issue commands to the servos or other output devices to make reality follow the plan. The desired position of the robot or one of its servos is often referred to as the *rabbit* and the position of the actual robot or servo is referred to as the *hound*. This terminology comes from the dog racing business, where a mechanical rabbit is driven down the track ahead of the pack of racing canines. A *closed loop* system is one that measures the error between the rabbit and hound, and attempts to minimize the gap without becoming erratic.

Classical control theory has been in existence for a very long time. Engineering courses have traditionally taught mathematical techniques involving poles and zeros and very abstract mathematics that beautifully describe the conditions under which a control can be designed with optimal performance. Normally, this involves producing a control system that maintains the fastest possible response to changing input functions without becoming underdamped (getting the jitters). It is perhaps fortunate that the professor who taught me this discipline has long ago passed from this plane of existence, else he would most probably be induced to cause me to do so were he to read what I am about to say.

Control theory, as it was taught to me, obsessed over the response of a system to a step input. Creating a control that can respond to such an input quickly and without overshooting is indeed difficult. Looking back, I am reminded of the old joke where the patient lifts his arm over his head and complains to the doctor "It hurts when I do this, Doc," to which the doctor replies, "Then don't do that."

With software controls, the first rule is never to command a control to move in a way it clearly cannot track.

If you studied calculus, you know that position is the integral of velocity, velocity is the integral of acceleration, and acceleration is the integral of jerk. Calculus involves determining the effect of these parameters upon each other at some future time.

In real-time controls this type of prediction is not generally needed. Instead, we simply divide time into small chunks, accumulating the jerk to provide the acceleration, and accumulating the acceleration to provide the velocity, etc. This set of calculations moves the rabbit, and if all of these parameters are kept within the capability of the hardware, a control can be produced that will track the rabbit.

Furthermore, the operator who controls the "rabbit" at a dog race makes sure that the hounds cannot pass the rabbit, and that it never gets too far ahead of them. Similarly, our software rabbit should adjust its acceleration if the servo lags too far behind.

A common problem with applying standard control theory is that the required parameters are often either unknown at design time, or are subject to change during operation. For example, the inertia of a robot as seen at the drive motor has many components. These might include the rotational inertia of the motor's rotor, the inertia of gears and shafts, rotational inertia of its tires, the robot's empty weight, and its payload. Worse yet, there are elements between these components such as bearings, shafts and belts that may have spring constants and friction loads.

Not withstanding the wondrous advances in CAD (computer aided design) systems, dynamically modeling such highly complex systems reliably is often impractical because of the many variables and unknowns. For this reason, when most engineers are confronted with such a task, they seek to find simpler techniques.

Flashback...

As a newly minted engineer, my first design assignment was a small box that mounted above the instrument panel of a carrier borne fighter plane. The box had one needle and three indicator lights that informed the pilot to pull the nose up or push it down as the multi-million dollar aircraft hurtled toward the steel mass of the ship. As I stood with all of the composure of a deer caught in the landing lights of an onrushing fighter, I was told that it should be a good project to "cut my teeth on!"

The design specification consisted of about ten pages of Laplace transforms that described the way the control should respond to its inputs, which included angle-of-attack, air speed, vertical speed, throttle, etc. Since microprocessors did not yet exist, such computers were implemented using analog amplifiers, capacitors, and resistors, all configured to perform functions such as integration, differentiation, and filtering. Field effect switches were used to modify the signal processing in accordance to binary inputs such as "gear-down."

My task was simply to produce a circuit that could generate the desired mathematical function. The problem was that the number of stages of amplifiers required to produce the huge function was so great that it would have been impossible to package the correct mathematical model into the tiny space available. Worse yet, the accumulation of tolerances and temperature coefficients through so many stages meant the design had no chance of being reproducible. With my tail between my legs, I sought the help of some of the older engineers.

To my distress, I came to realize that I had been given the task because, as a new graduate, I could still remember how to manipulate such equations. I quickly came to the realization that none of the senior engineers had used such formalities in many years. They designed more by something I can best describe as *enlightened instinct*. As such, they had somehow reprogrammed parts of their brains to envision the control process in a way that they could not describe on paper!

Having failed miserably to find a way out, I took a breadboarding system to the simulator room where I could excite my design with the inputs from a huge analog simulator of the aircraft. I threw out all the stages that appeared to be relatively less significant, and eventually discovered a configuration with a mere three operational amplifiers that could produce output traces nearly identical to those desired. It passed initial testing, but my relief was short lived. My next assignment was to produce a mathematical report for the Navy showing how the configuration accomplished the desired function!

At this point I reverted to the techniques I had honed in engineering school during laboratory exercises. I fudged. I manipulated the equations of the specification and those of the desired results to attempt to show that the difference (caused by my removal of 80% of the circuitry) was mathematically insignificant. In this I completely failed, so when I had the two ends stretched as close to each other as possible, I added the phrase "and thus we see that" for the missing step.

A few months later my report was returned as "rejected." Knowing I had been exposed, I began planning for a new carrier in telemarketing. I opened the document with great apprehension. I found that the objections had to do with insufficient margins and improper paragraph structures. I was, however, commended for my clear and excellent mathematical explanation!

Basic PID controls

At the simple extreme of the world of controls is the humble PID control. PID stands for Proportional, Integral, and Derivative. It is a purely *reactive* control as it only responds to the system error (the gap between the rabbit and the hound). PID controls were first introduced for industrial functions such as closed loop temperature control, and were designed using analog components. Amazingly, to this day, there are still a very significant number of analog PID controls sold every year. Because of its simplicity, many robot designers have adapted software PID controls for servo motor applications. Figure 5.1 shows a simplified block diagram of a PID controller. Since some of the terms of a PID do not work optimally in motor controls, we will consider the classical examples of PIDs in temperature control, and then work our way toward a configuration more capable of driving motors.

The PID control is actually three separate controls whose outputs are summed to determine a single output signal. The current reading of the parameter being controlled is subtracted from the command (in this case a set point potentiometer) to generate an *error* signal. This signal is then presented to three signal processors.

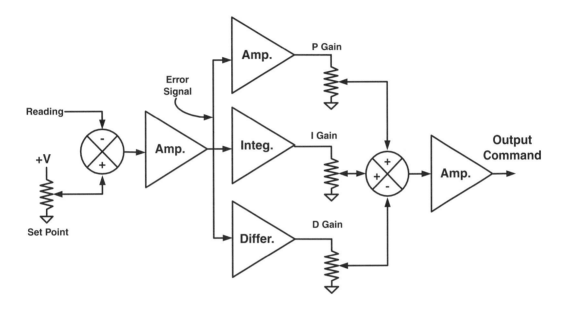

Figure 5.1. A classic analog PID control

The error proportional term

The first processor is a straight amplifier. This "proportional" stage could be used by itself but for one small problem. As the reading approaches the set point, the error signal approaches zero. Therefore, a straight proportional amplifier can never close to the full command.

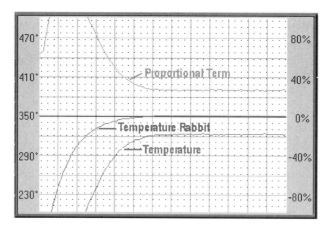

Figure 5.2. Critically damped proportional control with classic P-droop

Remember that there is a lag between the time the control generates an output and the reading reflects it. Therefore, if the gain of a proportional stage is increased too much in an attempt to minimize the error, the system will begin to oscillate. The minimum obtainable steady-state error that can be achieved is called the *P Droop* (see Figure 5.2).

The error integral term

What is needed at this point is a way of eliminating the P Droop. By placing an integrator in the loop, the system can slowly but continuously increase or decrease the output as long as there is any error at all. When the error is eliminated the integral term stops changing. At steady state, the error will go to zero and the integral term will thus replace the proportional term completely. Zero error may sound good, but remember it took quite some time to achieve it.

Since the integration process adds an additional time delay to the loop, if its gain is too high it is prone to induce a lower frequency oscillation than that of the *proportional-term* (P-term). For this reason, the P-term is used to provide short-term response, while the I-term (integral-term) provides long term accuracy.

The Achilles heel of the *integral term* is something called *integral wind-up*. During long excursions of the set point command, the integral term will tend to accumulate an output level that it will not need at quiescence. As the set point is approached by the reading, it will tend to overshoot because the algorithm cannot quickly dump the integral it has accumulated during the excursion. This problem is referred to as *integral wind-up*. Generally, the integral term is better suited to temperature controls than to motor controls, especially in mobile robots where the servo target is constantly being revised.

In motor control applications the speed command often varies so rapidly that the integral term is of little value in closing the error. For this reason, I do not recommend the integral term in any control where the power will normally go to zero when the rabbit reaches its destination (such as a position commanded servo on a flat surface).

In applications where an integral is appropriate, and where the rabbit tends to stay at set point for extended periods, an *integral hold off* is usually helpful. This simple bit of logic watches the rabbit and when its absolute derivative becomes small (the rabbit isn't changing much) it enables the integral term to accumulate. When the rabbit is in transition, the integral is held constant or bled off.

Generally, introducing sudden changes into the calculation of the PID (such as dumping the integral) is not an optimal solution because the effects of such excursions tend to destabilize the system.

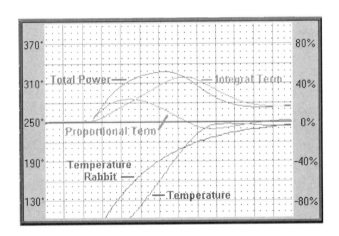

Figure 5.3. A well-tuned PID using only proportional and integral terms

Figure 5.3 shows a well-tuned PID control using only these P and I terms. The total power, proportional, and integral terms correspond to the scale on the right side of the graph, while the scale for the rabbit and temperature are on the right.

Notice that the temperature rabbit is an exponential function that approaches the set point of 250 degrees. The trick here is that while the temperature clearly overshoots the rabbit, it almost perfectly approaches the desired quiescent state. Over-shooting the rabbit causes the proportional term to go negative, decelerating the approach to set point. The integral term "winds up" during the ascent of the rabbit, but is cancelled out by the negative excursion of the proportional term as the temperature overshoots the rabbit. There is no *integral hold-off* in this example.

So why would we want anything more than this? The answer is that the performance is only optimal for the precise rabbit excursion to which the gains were tuned. Had we elected to send the rabbit to 350 degrees with these same gains, there would have been a significant overshoot because the integral would have wound up excessively. Since robot servos must operate over a wide range of excursions and rates, a more robust control is very desirable.

The error derivative term

With the tendency of the P-term and I-term to cause oscillations, it is natural to look for a counterbalance. The purpose of the D-term is just that. The D-term is a signal that is proportional to the derivative or rate-of-change of the error. This term is *subtracted* from the other terms, in order to suppress changes in the error. Thus, the whole PID equation that must be coded is merely:

$$P_t = (K_p \cdot E) + (K_i \cdot \textstyle\int E) - (K_d \cdot dE/dt)$$

Where P_t is the total power command, E is the current error, and K_p, K_i, and K_d are the proportional, integral, and derivative gains. The algorithm is implemented by repeatedly calling a subroutine on a timed basis. The frequency of execution depends on the time frame of the response. The higher the frequency, the faster the servo can respond, but the more computer resources that are absorbed. Generally the *PID rate* for a mobile robot operating at less than 3 mph will be from 10 Hz to 100 Hz. A single execution of the PID is often called a *tick* (as in the tick of a clock).

The proportional term is a simple, instantaneous calculation. The integral term, however, is produced by multiplying the current error by the integral gain and adding the result to an accumulator. If the error is negative, the affect is to subtract from the accumulator.

The derivative can be calculated by simply subtracting the error that was present during the previous execution of the algorithm from the current error. In some cases, this may be a bit noisy and it may be preferable to average the current calculation with the past few calculations on a weighted basis.

Unfortunately, the D-term can also induce instability if its effective gain is greater than unity, and it amplifies system noise. I have found this term to be of some limited value in temperature control applications, but of much less value at all in controlling servo motors. In fact, purely *reactive* controls such as PIDs tend in general to be of limited use with most servo motors.

Predictive controls

The *reactive* controls we have been discussing are driven entirely by error. This is a bit like driving a car while looking out the back window. Your mistakes may be obvious, but their realization may come a bit too late. By running controls strictly from the error signal, we are conceding that an error must occur. Indeed, there will always be a bit of error, but wouldn't it be nice if we could guess the required power first, and only use the remaining error to make up for relatively smaller inaccuracies in our guess?

Predictive controls do just this. Predictive controls do not care about the error, but simply watch the rabbit and try to predict the amount of power required to make the servo track it. Since they watch the rabbit, and are not in the feedback loop, they are both faster and more stable than reactive controls.

The rabbit term

Looking at the temperature rabbit curve of Figure 5.3, we begin to realize that there are two relationships between the rabbit and the required power that can be readily predicted. The first relationship is that for any steady-state temperature, there will be a certain constant power required to overcome heat loss to the environment and maintain that temperature. Let's call this the *rabbit term*, and in the simplest case it is the product of the rabbit and the *rabbit gain*. For a temperature control this is the amount of power required to maintain any given temperature relative to ambient. For the drive motor of a robot, this will be the amount of power that is required to overcome drag and maintain a fixed speed.

The power required may not bear a perfectly linear relationship with the set point over its entire range. This *rabbit gain* will vary with factors such as ambient temperature, but if the operating temperature is several hundred degrees, and the ambient variation is only say 20 degrees, then this power relationship can be assumed to be a constant. If the ambient temperature varies more appreciably, we could take it into account in setting the gain. *Note that in some cases the relationship may be too nonlinear to get away with a simple gain multiplier. In these cases an interpolated look-up table or more complex mathematical relationship may be useful.*

The rabbit derivative and second derivative terms

The second relationship we can quickly identify has to do with the slope or derivative of the rabbit. For a position servo, the rabbit derivative is the *rabbit velocity,* and the rabbit 2nd derivative is the *rabbit acceleration.*

If the rabbit is attacking positively, we will need to pour on the coals to heat the furnace up or to accelerate the robot. This term is thus proportional to the derivative of the rabbit, and we will call its gain the *rabbit derivative gain* or *rabbit velocity gain.*

If the rabbit is accelerating, then we will need even more power. This term is most commonly found in position controls and its gain is thus called the *rabbit acceleration gain.* If we perfectly guess these gains, then the servo will track the rabbit perfectly, with the minimum possible delay.

The big advantage to the rabbit derivative terms as opposed to the error derivative term is that, because they work from the nice clean rabbit command, they do not amplify system response noise.

Combined reactive and predictive controls

Unfortunately, a predictive control of this type has no way of correcting for variables such as ambient temperature or payload. It is also unlikely that our perfectly selected gains will be as perfect for every unit we build. It is possible to learn these gains during operation, but this will be discussed later.

It is therefore not practical to expect exact performance from a predictive control. Yet, the prediction may buy us a guess within say 10% of the correct value, and its output is instantaneous and stable. For this reason, it is often optimal to combine terms of a conventional PID with those of a predictive control. Using our analog control as a metaphor, consider the configuration shown in Figure 5.4.

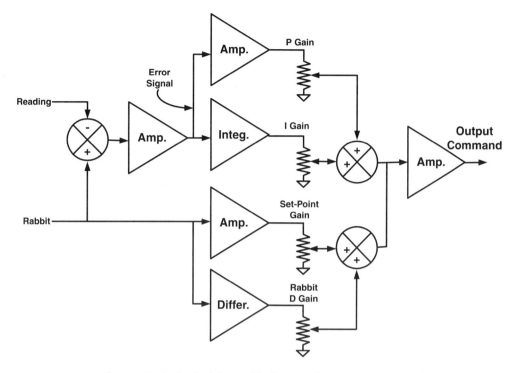

Figure 5.4. Hybrid predictive and reactive control

In this configuration, the instantaneous rabbit value is multiplied by a gain to produce the *rabbit term*. This term is sometimes called the *set point gain* because at set point (quiescence) this term should produce the bulk of the output needed.

The second new function is a differentiator that provides a term proportional to the rate of change of the rabbit. When added to the *rabbit term*, this term gives us a much better guess at the power required during transitions of the rabbit. Additionally, we have retained the *proportional* and *integral* terms driven by the error to make up for any inaccuracy in the prediction. The result is an extremely capable yet simple control architecture suitable for a wide range of applications.

Various PID enhancements

Since the standard PID control algorithm is far from ideal for many types of controls, software engineers have produced an almost endless array of variations on the basic theme. In most cases, these modifications amount to replacing the gain constants we have been discussing with gain tables or simple functions.

Asymmetric gains

One of the most common problems encountered is that many systems are asymmetric in their response to a power command. For example, a heater may be adequately powerful to quickly increase the temperature of a mass with say 30% power, but when the same amount of power is removed, the temperature may not drop nearly as fast as it had increased. This asymmetric response is obviously due to the fact that cooling is not merely the result of removing power, but rather the result of air convection.

The drive power requirement for a heavy robot is also very asymmetric. While a good deal of forward power is usually required to accelerate the robot, only a small amount of reverse power is required to decelerate it at the same rate. If a symmetric PID is used to control such a motor, it may be impossible to obtain adequate forward responsiveness without causing the robot to summersault when deceleration occurs. Things get even more exciting when a robot goes over the top of a ramp and starts down the back side.

For asymmetric loads, it is often useful to provide two gains for each term. One gain is used for positive term inputs, while the other is used for negative inputs.

Error band limits

Another useful modification to reactive PID terms is to limit the error range over which they respond proportionally. If the error is within this band, then the error is multiplied by the gain, otherwise the appropriate error limit is used. This is particularly useful with the integral term to prevent wind-up. The example below is coded in Visual Basic, and utilizes many of the techniques discussed thus far.

In all these cases, you will be governed by the principal of *enlightened instinct*. To become a Zen master of such techniques, you must understand the causes of problems, and the basics of physics and dynamics, but in the end the solution is often half theory and half instinct. If the choice is made to be rigidly mathematical in the solution, you may still end up guessing at a lot of parameters, and producing an algorithm that sucks every available cycle out of your control computer.

My personal preference for creating such controls is to design the software so that virtually every factor can be changed on the fly during operation. You can then run the system and observe its response to changes in various parameters. For example, the robot can be set to drive backward and forward between two points while various gains are tried. *In other words, let the system talk to you.*

Figure 5.5 shows typical code for a thermal PID control using many of the terms and tricks just discussed.

```
'Calculate the PID for a single control using error proportional,
'error integral, rabbit, and rabbit derivative terms. Error and
'rabbit derivative gains are non-symmetric.

'The process is controlled by an array of singles called
'"ControlSingles". Each term also has a limit band on its error.
'If the error is greater than the band, then the band value is
'substituted for the limit.

'Routine returns the power command as an integer between 0 and 9999.
Public Static Function DoPIDs(TempRabbit As Single, Temp As Single) As Integer

Dim Error As Single      'Raw error
Dim LimError As Single  'Error or limit, whichever is smaller.
Dim OutputAccum As Long
Dim PGain As Long   'Proportional command Gain (0 to 9999)
Dim DGain As Long   'Derivative command Gain
Dim IGain As Long   'Integral command Gain
Dim RGain As Long   'Setpoint rabbit command Gain
Dim PTerm As Single
Dim ITerm As Single
Dim DTerm As Single
Dim RTerm As Single
Dim RabbitDeriv As Single
Dim LastTempRabbit As Single
Dim IntegralHold(MaxZones) As Integer '0=No Hold, 1=allow pos. only, -1=allow neg.

    On Error Resume Next

    'Calculate the error and rabbit derivative.
    Error = TempRabbit - Temp
    RabbitDeriv = TempRabbit - LastTempRabbit
    LastTempRabbit = TempRabbit

    'Get the Rabbit and error gains
    RGain = ControlSingles(RabbitGain)  'Rabbit gain is always positive.

    'Some gains depend on error polarity
    If Error >= 0 Then 'For positive errors use positive gains
       PGain = ControlSingles(PposGain)
       IGain = ControlSingles(IposGain)
    Else
       PGain = ControlSingles(PnegGain)
       IGain = ControlSingles(InegGain)
    End If
```

```
'
'Since there is no error derivative term, we will use
'Dgain to mean the rabbit derivative gain. Its gain
'depends on the polarity of the rabbit derivative.
If RabbitDeriv > 0 Then
    DGain = ControlSingles(DposGain)
Else
    DGain = ControlSingles(DnegGain)
End If

'Now do the calculation for each term.
'First limit the error to the band for each gain
If Error > ControlSingles(PBand) Then
    LimError = ControlSingles(PBand)
ElseIf Error < -ControlSingles(PBand) Then
    LimError = -ControlSingles(PBand)
Else
    LimError = Error
End If

PTerm = CDbl((PGain * LimError) / 100)

If Error > ControlSingles(IBand) Then
    LimError = ControlSingles(IBand)
ElseIf Error < -ControlSingles(IBand) Then
    LimError = -ControlSingles(IBand)
Else
    LimError = Error
End If

'The I term is cumulative, so it's gain range is 1/100th that of P.
'Integral is bled off while rabbit is moving, or if the output
'accumulator has gone below zero with a negative integral or over
'full range with a positive integral.

If Abs(RabbitDeriv) < ControlSingles(AttackRate) / 30 Then
    If (LimError > 0 And IntegralHold >= 0) Or _
       (LimError < 0 And IntegralHold <= 0) Then
        ITerm = LimitTerm(ITerm + ((IGain * LimError) / 10000))
    End If
Else 'Bleed off the i term.
    ITerm = 0.99 * ITerm
End If

DTerm = LimitTerm(RabbitDeriv * DGain * 10)
RTerm = LimitTerm(TempRabbit * (RGain) / 500)
OutputAccum = PTerm + ITerm + DTerm + RTerm
```

```
'Limit the output accumulator and flag the integrator
'if the output goes out of range. (In a properly tuned
'control, this should never happen).
If OutputAccum > 9999 Then
   OutputAccum = 9999
   IntegralHold = -1 'Allow only down integration
ElseIf OutputAccum < 0 Then
   OutputAccum = 0
   IntegralHold = 1 'Allow only upward integration
Else
   IntegralHold = 0 'Allow both directions.
End If

   DoPIDs = CInt(OutputAccum)

End Function


'------------------------------------------------------------------------
'Prevent Terms from exceeding integer range for logging.
Private Function LimitTerm(Term As Single)
   If Term > 9999 Then
      LimitTerm = 9999
   ElseIf Term < -9999 Then
      LimitTerm = -9999
   Else
      LimitTerm = Term
   End If
End Function
```

Figure 5.5. A band-limited hybrid control with integral hold-off

Robot drive controls

The drive system of a mobile robot is subject to widely varying conditions, and thus it presents several challenges not found in the simpler control systems already discussed. These include:

1. Drive/brake asymmetry

2. Quadrant crossing nonlinearity due to gear backlash

3. Variable acceleration profiles due to unexpected obstacle detection

4. Variable drag due to surface variations

5. Variable inertia and drag due to payload variations.

Drive-brake asymmetry

There are two basic categories of drive gear reduction: back drivable and non-back drivable. The first category includes pinion, chain, and belt reducers, while the second includes worm and screw reducers. Servos using non-back drivable reducers are easier to control because variations in the load are largely isolated from the motor. Back-drivable reducers are much more efficient, but more difficult to control because load fluctuations and inertial forces are transmitted back to influence the motor's position.

The response of either type servo is usually very nonlinear with respect to driving and braking. To achieve a given acceleration may require very significant motor current, while achieving the same magnitude of deceleration may require only a minute reverse current. For this reason, asymmetric gains are usually required in both cases.

Quadrant crossing nonlinearity

Most drives are either two or four quadrant. A two-quadrant drive can drive in one direction and brake. A four-quadrant drive can drive and brake in both directions. The term quadrant thus comes from these four possible actions of the drive. A nasty phenomenon that appears in the back drivable servos is that of nonlinear regions in a servo's load. A drive motor using a pinion gear box is an excellent example. When the power transitions from forward to reverse (called *changing quadrants*), the servo will cross a zero load region caused by the gear backlash. If the control does not take this into account, it may become unstable after freewheeling across this region and slamming into the opposite gear face. One trick for handling this discontinuity is to sense that the control has crossed quadrants and reduce its total gain after the crossing. This gain can then be smoothly restored over several ticks if the control remains in the new quadrant.

Natural deceleration

Any servo will have its own natural deceleration rate. This is the deceleration rate at which the servo will slow if power is removed. If the programmed deceleration is lower than the natural deceleration, the servo never goes into braking and the response is usually smooth.

Thus, during normal operation the ugly affects of quadrant crossing may not be noticed. If, however, the robot must decelerate unexpectedly to avoid a collision, the servo may react violently, and/or the robot may tip over forward. For this reason, a maximum deceleration is usually specified, beyond which the robot will not attempt to brake. In this way the robot will slow as much as possible before the impact. If a mechanical bumper is designed into the front of the robot, it can be made to absorb any minor impact that cannot be avoided by the control system.

Freewheeling

Perhaps the most difficult control problem for a drive servo is that of going down a ramp. Any back drivable drive servo will exhibit a freewheeling velocity on a given ramp. This is the speed at which the robot will roll down the ramp in an unpowered state. At this speed, the surface drag and internal drag of the servo are equal to the gravitational force multiplied by the sine of the slope. The freewheeling speed is thus load dependent.

If a robot attempts to go down a ramp at a speed that is greater than its natural free-wheeling speed for the given slope, then the servo will remain in the forward driving quadrant. If the robot attempts to go slower than the freewheeling speed, then the servo will remain in the braking quadrant. The problem comes when the speed goes between these two conditions. This condition usually occurs as soon as the robot moves over the crest of the ramp and needs to brake.

Under such transitions, both the quadrant discontinuity and drive/brake nonlinearity will act on the servo. This combination will make it very difficult to achieve smooth control, and the robot will lurch. Since lurching will throw the robot back and forth between driving and braking, the instability will often persist. The result roughly simulates an amphetamine junky after enjoying a double espresso. If the gain ramping trick described above is not adequate, then it may be necessary to brake.

My dearly departed mother endeared herself to her auto mechanic by driving with one foot on the gas and the other on the brake. When she wished to go faster or slower she simply let up on the one while pushing harder on the other. This method of control, however ill-advised for an automobile, is one way of a robot maintaining smooth control while driving down ramps.

One sure-fire method of achieving smooth control on down ramps is to intentionally decrease the freewheeling velocity so that the servo remains in the drive quadrant. To accomplish this, one can use a mechanical brake or an electrical brake. The simplest electrical brake for a permanent magnet motor is to simply place a low value *braking resistor* across the armature. While a braking resistor or other form of braking will reduce the freewheeling speed of the robot, it will waste power. For this reason, brakes of any sort must be applied only when needed.

The ideal way to reduce the freewheeling velocity of a drive servo is through the use of circuitry that directs the back EMF of the motor into the battery. In this way, the battery recovers some energy while the robot is braking. The common way of doing this is through the use of a *freewheeling* diode in the motor control bridge.

I have not found simple freewheeling diodes to provide an adequate amount of braking in most downhill situations. This is because the back EMF must be greater than the battery voltage and the effective braking resistance includes both the battery internal resistance and the motor resistance. Thus, voltage-multiplying circuits are usually required if this type of braking is to actually be accomplished.

Drag variations

A robot drive or steering servo is usually position commanded. The moment-to-moment position error of a cruising mobile robot is usually not critical. As long as the robot's odometry accurately registers the movements that are actually made, the position error can be corrected over the long term. For this reason, drag variations are not usually critical to normal driving.

However, sometimes a robot will be required to stop at a very accurate end position or heading. One such case might be when docking to a charger. In these cases, the *closing error* becomes critical.

Variations in drag caused by different surfaces and payloads can also cause another significant problem. If a robot's drive system is tuned for a hard surface, and it finds itself on a soft surface, then the servos will not have adequate gain to close as accurately.

If the robot stops on such a surface with its drive servo energized, then the position error may cause the servo to remain stopped but with a significant current being applied to the motor. This type of situation wastes battery power, and can overheat the motor. If the robot's gains are increased for such a surface, it may exhibit instability when it returns to a hard surface.

One solution to this situation is to ramp up the *error proportional* gain of the servo as its speed command reaches zero. When the P-gain has been at its upper limit (typically an order of magnitude higher than its base value) and long enough for the closing error to be minimized, the gain is smoothly reduced back to its base setting. I have found this technique quite effective.

Tuning controls

The problem that now arises is one of tuning. There are many terms that all interplay. The first thing that is needed is a good diagnostic display upon which we can view the rabbit, reading, power output, and various term contributions. The graphs shown in this chapter are from an actual temperature control system written in Visual Basic. Readings are stored into a large array, and then displayed on the graphs in the scale and format desired.

Figure 5.6 shows the performance that can be obtained from a control like that shown in Figure 5.5. Notice this control cleanly tracks the rabbit by largely predicting the amount of power required. As the temperature rabbit is climbing linearly, the *rabbit derivative term* is constant, and as the rabbit begins to roll into the set point, this term backs off its contribution to the total power.

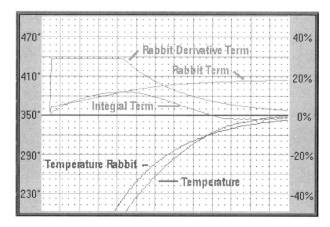

Figure 5.6. Combination predictive and reactive control

The rabbit term, on the other hand, represents the power required to remain at the present rabbit temperature. Thus, this term is proportional to the value of the rabbit. For clarity, the *error proportional term* is not shown, but it would have looked like that in Figure 5.3, reversing polarity as the temperature reading crossed the rabbit. The double effect of the *rabbit derivative term* and the *error proportional term* helped prevent overshooting of the set point.

Tuning such controls is an iterative process. Normally, the control is first run with all gains but one or two set to zero.

Learning gains

For systems with an *error integral term*, it is a fairly simple matter to write code that can learn the rabbit gain just discussed. The process is one of finding reasonable parameters for the integral and proportional gains, and then allowing the system to reach a quiescent state in the region of typical operation.

The integral term should come to replace the proportional term, and at this point, it is divided by the set point to provide the *rabbit gain*. Since this will replace the power previously being supplied by the error integral term, the integral accumulator is dumped, and the new *rabbit term* generates the entire output required.

If the rabbit gain was already nonzero, then the old rabbit term must be added to the error integral term and then the result divided by the set point to calculate a corrected rabbit gain.

The *rabbit derivative gain* can then be determined by turning all gains to zero except the rabbit gain. The rabbit is then commanded to begin a smooth transition and the rabbit derivative gain is trimmed to make the slope of the reading match the slope of the rabbit.

Tuning robot position controls

In the case of motor controls, the simple gain learning processes previously described may not be practical. It is theoretically possible to write adaptive controls that can adjust their own parameters dynamically, but such techniques are well beyond the scope of this chapter. The simplest method is the remote tuning process.

Motor controls may be velocity commanded, but in robots they are more commonly position commanded. The velocity desired is imparted through the motion of the rabbit. In most cases, absolute velocity accuracy is not as important as smoothness.

A position-seeking control will normally consist of a *rabbit velocity term*, a *rabbit derivative (acceleration) term*, an *error proportional term*, and a less significant *error derivative term*. Since no power is required to hold a robot at the same place (on a flat surface), there is no need for a *rabbit term*. On a slope, the *error proportional term* should be adequate to prevent the robot from rolling once it has stopped. This implies that the power will remain on to the control when the rabbit has stopped, and this brings about other issues.

Rabbits chasing rabbits

The smoother the rabbit's motion, the more likely the servo will be able to track it closely. For this reason, linear approximations (trapezoidal profiles) are not generally acceptable. Such a profile demands infinite acceleration at the knee points. Command curves with low second derivatives (jerk term) are also essential in some cases.

When we ascend to the level of making a robot navigate, there is much more involved than a simple single-axis servo chasing its rabbit. We must first generate a position rabbit that runs through imaginary space, and then determine the behavior we want from the steering and/or drive motors in order to follow it. If we are executing an arc, then the steering rabbit must be slaved to the drive rabbit, and both must be slaved to the position rabbit.

We will discuss some of these issues in the following chapters. The important thing to remember is that the most elaborate building is made of lots of simple blocks. Likewise, all complex systems are constructed from simpler blocks. If these blocks are understood and perfected, they can be used to create systems that dependably generate very complex behavior.

Conclusions

The terms of the hybrid reactive and predictive control we have discussed are shown in the table below, along with a matrix for the applications in which they may be useful.

Term	*Calculated From*	*Temperature Control*	*Position Control*
Error proportional	Rabbit – reading (error)	√	√
Error derivative	Rate of change of error	√	√
Error integral	Accumulation of error	√	
Rabbit	Rabbit value (temp or position)	√	
Rabbit derivative	Rabbit rate of change (velocity)	√	√
Rabbit 2nd derivative	Rabbit velocity rate of change (acceleration)	√	√

The subject of controls is vast, and there are many techniques more advanced than the ones described here. Work has even been done with neural networks to solve the thornier problems. Even so, the simple architecture described here is one that I have found to be very useful, and it should serve as a starting point. More importantly, it demonstrates a way of thinking through such problems logically. In the chapters ahead, we will build other strategies using the same logical processes.

CHAPTER **6**

Communications and Control

In mobile robot design, there are few decisions more critical than deciding how communications will be performed. The method in which the data is encoded in a message is called the *protocol*. There are thousands of different protocols floating around the world. Some protocols are concerned only with piping data from one place to another without regard to what it means, while other protocols are concerned with actual meaning of the data.

Some protocols have *layers* for each of these purposes and more. A protocol or layer of a protocol that concerns itself with the meaning of the data is usually called the *application layer* or *application protocol*. This is the protocol we need to decide upon. While this protocol may ride on other protocols, it is still the native interface language of the machine.

As an example, on the internet the most common application protocol for browsers is *HTML*. This *hypertext markup language* acts to convey the way a page should appear and respond to the recipient's mouse actions. The actual transmission of HTML is accomplished by other protocols such as TCP. We need a similar, and equally flexible, protocol that meets the needs of controlling and monitoring our machine.

Before getting into the requirements of a robot's application protocol, it is useful to briefly consider the network technology that will be carrying our messages.

Popular networks

In the past, the natural choice for inter-robot communications was ASCII, commonly at RS-232, RS-485, and RS-422 levels. ASCII was designed for point-to-point communications, but could be easily adapted to master-slave networking.

ASCII has come to be called *serial* communications although virtually all communications are now serial. The reason for ASCII's popularity was simply that these interfaces were available for all microprocessors and were built directly into many CPU chips.

The biggest disadvantage of ASCII is that it is limited to baud rates of 115 Kbps. The proliferation of intelligent controls and the internet have brought into existence several new and better alternatives to serial communications. These networks have both electrical and protocol definitions on how data is transmitted, but they do not specify what the data means. The meaning of the data is determined by the applications protocol that rides on the network protocol.

USB and FireWire

Both USB (Universal Serial Bus) and FireWire (IEEE-1394) were designed to serve as peripheral busses for PCs. They are therefore inherently master-slave in nature. The USB interface is commonly used in low-to-medium bandwidth peripheral connections, ranging from image scanners to digital cameras. USB operates up to 1.25 Mbps.

FireWire is more expensive than USB, but much faster. Offering data rates up to 400 Mbps, it is most often found in computers that are designed specifically for high-end multimedia editing.

Either of these busses can serve as a data acquisition bus on board a mobile robot, but they are less suited for higher level communications.

CAN

The CAN (Controller Area Network) is a multi-master asynchronous protocol that was developed for the automotive industry. Over the past decade it has come to dominate inter-vehicle communications. In most modern automobiles, almost every electrical object in the vehicle from the ignition to the automatic windows communicates by CAN network. There is wide support for interfacing to various instruments and development software is available at a low cost. Manufacturers such as Motorola and National Instruments offer chip sets and extensive support for CAN. The presence of the CAN network has made possible diagnostics that have revolutionized automotive repair and maintenance.

There are two electrical standards for CAN. The low speed (125 Kbps) standard trades off bandwidth for the ability to continue operating if one of the two lines of the communications cable becomes shorted to ground. The high-speed version can operate up to 1 Mbps, but is not as robust. Both standards are often used in different areas of a single automobile.

CAN has the advantage of being developed especially for the automotive industry, but it does not have the wide support outside this application arena that Ethernet enjoys.

Ethernet

The Ethernet protocol has become widespread and is probably the most popular protocol for new robotics projects. There are several electrical standards for coax, twisted pair, and even fiber-optic connectivity. Ethernet is gradually replacing dedicated protocols in applications such as security monitoring and building controls. Simple interfaces are available to rates of 100 Mbps. The availability of low cost wireless Ethernet (802.11) make Ethernet almost ideal for mobile robotics.

The biggest disadvantage to Ethernet is that it was not designed as a control or data acquisition protocol, and it offers limited control over message latency. Even so, with proper time stamping of data, Ethernet can be used for most of the data acquisition and control functions a mobile robot will need to perform.

In the final event, most robots will contain a combination of these networks. For example, a lidar may have a serial interface that is controlled by a sub-processor that in turn communicates via Ethernet to a higher-level computer. This translation may be dictated by the fact that the lidar is only available with a serial interface. Even if the lidar has an Ethernet interface, it will undoubtedly have its own application protocol.

Unfortunately, the necessity of changing protocols causes road blocks to communications across these boundaries. Only requests and commands supported at both sides of the interface will be available through it.

Basic requirements of the application protocol

It is absolutely essential that the application protocol for a mobile robot support the monitoring and manipulation of almost every function of the robot, and that it do so efficiently. This is true even if the robot is not expected to communicate with a base station during normal operation. It is even true if the final design will have no means of communications at all! I have watched in horror as engineers doomed their projects with ill-advised protocols.

There are three primary reasons why the protocol must be flexible. First, it must support debugging of onboard software; secondly, it must facilitate future expansion; and finally, it must facilitate the coordination of the robot with other systems such as elevators, gates, and so forth. Let's consider these reasons in order.

Support of remote diagnostics

Mobile robots interact with their environments in ways that are virtually nondeterministic (see more about this in Chapter 17). A robot will interact with any given environment in an almost infinite variety of ways, and new environmental factors are often introduced to a robot long after its development has been "finalized." The result is that the robot's software will evolve indefinitely if it is a commercial success.

With stationary control systems, it is possible to be in front of the control computer as a system operates and to use any number of diagnostic tools to aid in debugging. If the program is written in a language such as C++ or VB, these tools are already built into the language. For PC-based systems, the movement toward network-enabled languages such as VBNet[1] also facilitate remote operations. If the system or subsystem is being programmed in assembly language, the tools are usually built into the development system or its in-circuit emulator[2]. When, in the course of operation of such a stationary system, any of these fixed computers must communicate with other systems, the protocol to do this is often designed after the basic system operation has been debugged.

For remote diagnostics in stationary PC-based systems, it may even be possible to install a dial-up modem or internet connection and to use off-the-shelf remote viewing programs such as pcAnywhere[3]. While extremely inefficient from a bandwidth standpoint, this approach may still be effective.

Mobile robots are a different matter. The two biggest differences are the fact that the robot system is mobile, and that it is usually composed of several separate subsystems. If

---

[1] VBNet is a trademark of Microsoft Corporation.

[2] An in-circuit emulator is a device that replaces the CPU chip of the target system with an umbilical to a console. The console allows emulation of the final system operation while providing debug capabilities.

[3] pcAnywhere is a trademark of Symantec Corporation.

the robot begins exhibiting strange behavior, it is essential for the programmer to be able to determine which sensor or algorithm has induced the behavior, and to accomplish this without walking along behind the robot. Additionally, it is sometimes necessary to trim the parameters of existing algorithms in order to adapt them to a particular environment (more about this in my next flashback).

Support of system evolution

The second important requirement for the communication protocol is that it must support the future evolution of the system. A commercial robotic system is like a living organism; it must adapt to take advantage of the opportunities and meet challenges in its application environment. Customers will want the system customized to accomplish things that were never envisioned during the design cycle. If doing this requires a change of any sort in the protocol itself, then the result will be a system that is not compatible with its siblings. When this happens, the various versions of software metastasize, and support becomes almost impossible.

Coordination of other resources

In many environments, it is necessary for a mobile robot to control fixed resources. For an indoor robot, these may include door openers and elevators. It is possible to put special systems on the robot such as short range garage-door controllers, but these solutions are often problematic because of the one-way nature of the control and the limited channel capacity.

A better approach in many cases is to allow the base station to control these resources for the robot(s). It is therefore important that the protocol be flexible enough that it support special flag exchanges to coordinate interfacing with these systems. The exact nature of such coordination may vary from application to application, but if the protocol is flexible this will not be a problem.

Rigid protocols and other really bad ideas

The most obvious (and common) solution to a protocol is to think of all of the things you will want to tell a robot and all of the things you will want it to tell you, and then simply to define a number of dedicated messages for exchanging this data.

Let's consider the case of a simple remote controlled robot with differential drive. Differential drive features one powered wheel on each side of the robot. When both wheels turn at the same speed, the robot moves forward. Steering is accomplished by moving one wheel faster than the other, or driving the wheels in opposite directions. We decide that all we need is to send this robot a command for the two motor speeds.

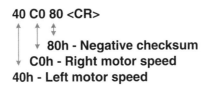

Figure 6.1. A simple, rigid protocol for a differential drive robot

The protocol in Figure 6.1 is about as simple as a protocol can get. Notice that two ASCII characters describe a hexadecimal 8-bit byte, and that we are using 2's complement for representation of the data. Thus, the above command says that the robot should run its left motor at half speed forward (positive) , and run its right motor at half speed in reverse (negative). In otherwords, we want the robot to turn in place. The two data bytes are subtracted from zero and the low byte of the result is sent as the checksum.

One of the important aspects of a protocol is *framing*. Framing is the means by which the receiver/decoder determines where a new message begins. Sending each byte as two ASCII characters is wasteful of bandwidth, but it allows us to be sure that the data will never cause the transmission of a byte having the value 0Dh. If this occurred, the data could be confused at the receiver with the terminating <CR> character. This could cause a framing error, and possibly even cause mistranslation of the data.

Let's assume we are happy with this protocol and begin shipping machines. What happens when we add a camera to the robot and want to be able to pan and tilt it? There is nothing in the above protocol that says what the message is about—we have just assumed since there was only one type of message the robot would know. But if we wish to add a pan and tilt message, the robot must be able to tell which commands are for which purpose.

We can combine the two messages into a single eleven character message or keep the messages separate and add a message type identifier to each. We realize that making our one message longer and longer will eventually lead to problems, so we decide on the second alternative as shown in Figure 6.2. Since we have already fielded robots with the first protocol, however, our new version of the base station control will not be compatible with our old robots, even if the old robots have no camera controls.

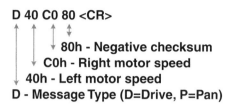

D 40 C0 80 <CR>

80h - Negative checksum
C0h - Right motor speed
40h - Left motor speed
D - Message Type (D=Drive, P=Pan)

Figure 6.2. Rigid protocol with message identifier

If we decide to keep the old message formats as we add new ones, things will eventually become very confusing. Worse, the receiving decoder will need to test incoming messages for all of these formats. This will make the decoding process absorb more system resources, and the possibilities for bugs abound. After years of patching such protocols, we will have created a nightmare. Worse yet, whenever we decide to read or set any parameter we had not previously included, we must replace the messaging program(s) at both ends of the communications link before the new data can be accessed.

The problem is that we cannot predict all of the messages we will eventually need for any but the simplest systems. To believe that it is possible to think of all the things you will want to communicate is a dangerous conceit. Within hours of establishing a link to the robot, you will find you need a bit more data, or you will realize that you need to influence the robot in an additional way. The protocol needs to be flexible enough to support the unexpected.

Flashback...

I am reminded of a very painful example. We were working on a project to adapt our commercial security robot for use in a military rack warehouse environment. A rack warehouse is one based on row after row of rack shelving. The problem was that the robot had been designed to operate in office environments. In its native habitat, it used sonar to look for walls and other flat surfaces from which it could correct its dead reckoning.

We were given access to a rack warehouse site where we could do testing. While racks could be at any height, there were always support posts on regular intervals, and we decided to base our approach on these. In the course of early development, we had used our limited funding to add some simple navigation tricks. These tricks used retro-reflective infrared sensors to detect strips of reflective tape mounted to the posts of the racks.

A week or so before a critical milestone test; we were informed that there had been a small administrative change. We would not be installing at the rack warehouse at all, but rather at a pallet warehouse full of crates. To us, the change was a bit like telling General Eisenhower on June 1st that the invasion of Europe was to be a land invasion through Russia instead of an amphibious assault at Normandy. What? Is there a problem?

As luck would have it, we had recently added control variables to the wall imaging algorithms that allowed us to navigate from the not-so-straight walls of office cubicles. Our flexible communications protocol and programming language allowed us, on site, to manipulate these variables and view the results. Thus, we were able to adapt the cubicle navigation variables to allow sonar navigation from crates.

This solution was not entirely kosher, since we had to move a significant number of crates to assure that they were roughly along the edges of the aisles. Even so, with a combination of flexible communication and control, ingenuity, and our newly acquired forklift driving skills we saved the day—or so we thought.

Although there were still several years of development time available before scheduled system deployment, several additional "administrative" changes would occur. Despite our warnings about the desperate need, funding for long-range navigation research was omitted from the next budget on the justification that the milestone test had demonstrated that the system needed no further navigational enhancement! Other participants, it seemed, needed the resources.

Our luck held out a bit longer, when in the nick of time we were able to fund the needed long range laser navigation from unrelated commercial contracts. Unfortunately, a final administrative change came when the RFP (Request for Proposals) was issued for the production of the system. The specification mandated that the communications protocol be changed from our flexible "proprietary" protocol to a rigid protocol that would be used by all future robots fielded by the program, both indoor and outdoor[4]. Through this and other key decisions, the program management snatched defeat from the very jaws of victory!

Rigid protocols are fairly common in peripheral systems such as printers, laser rangers, and other measurement systems. In fairness, a rigid protocol is often the appropriate choice in such systems. But even in these limited systems one often sees the beginnings of real problems for a rigid protocol.

---

[4] In this the program has thus far succeeded in that no robots of either type have yet been fielded. The contract to modify our robots was given to a giant defense contractor, and after several costly overruns, the warehouse robot program was placed in mothballs and the manager was given a promotion.

As an example, I recently interfaced a cable printer to a system and found that there were up to six different messages for performing each of the dozens of tasks. It was clear that some of these messages were more recent additions with added functionality, but the older messages were maintained for backward compatibility. For some reason, some of the older messages did not appear to work properly. The protocol was very inefficient and confusing, and it took an excessive effort to finally debug the interface.

Flexible protocols

For all of the reasons expounded upon, a robot communications protocol *must* be flexible. One way of making a protocol flexible is to allow it to access all public variables. This is called a *blackboard* approach. If access is to be restricted to certain areas, this can be accomplished in the encoder or decoder as opposed to making the protocol itself restrictive. This comes under the rule "You can always saw a board shorter, but you can't saw it longer."

Blackboards and exposing parameters with absolute addressing

In microcontroller applications, addressing is often absolute. Memory fields vary in size from a few kilobytes to a few megabytes. Thus, a memory location will always contain the same data and its address can be specified in a message to request or set that data.

The overhead of the message format will usually be such that it is very inefficient to request or set a single variable at a time. The simplest solution to this problem is to group variables together if they are likely to be written or read together. For example, the robot's azimuth, X position, and Y position will usually be requested together. If these variables are located contiguously, the protocol need only specify the beginning address of the block and the number of bytes.

More than two decades ago we were faced with the same requirements just discussed. We elected to use a protocol based on the Intel hex file format. This format described a method of placing an image of data in a file in a way that it could be read and loaded to absolute memory locations in a target system. This format was used most commonly for saving memory images to be used by a PROM (programmable read-only memory) burner. Our "set data" message was identical to a line of such a file. We used a spare data field to specify the destination computer, allowing the message to address up to 256 different blackboards or computers. Figure 6.3 shows a "set data" message in this format. The individual characters are in ASCII format, and as in the previous example, two characters are required to transmit each byte of data. This inefficiency was offset largely by the advantage that it could be monitored on a simple ASCII terminal, and by the security that this type of coding added.

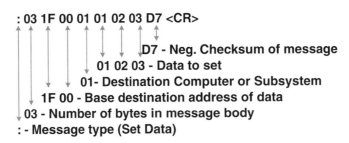

: 03 1F 00 01 01 02 03 D7 <CR>

D7 - Neg. Checksum of message
01 02 03 - Data to set
01- Destination Computer or Subsystem
1F 00 - Base destination address of data
03 - Number of bytes in message body
: - Message type (Set Data)

Figure 6.3. A blackboard set (write) data message

Upon accepting the message, the receiving (slave) process would return the checksum as a single binary byte. If the checksum answer were not received, the master would repeat the message.

The crude single-byte checksum would normally look like insufficient protection for a message that was to be transmitted over less than perfect means. Since the checksum could take on only 256 discrete values, it would seem that the protocol would miss detecting a multicharacter error once in 256 times. This is not the case. First, except for the message type, each character can take on only 16 values (0-9 and A-F). Thus, an error that occurs in a character transmission, in such a way that the result is no longer one of these characters, will be detected by the decoder before the checksum is ever considered. If a whole character were lost, the receiver would know that the message was not valid. Additionally, when the message was sent over a radio or modem link, it was always wrapped in a more secure protocol.

The message format for reading data is shown in Figure 6.4. A slave computer would respond to this message by returning the requested block of data, followed by the checksum of both the data and the requesting message. Response data was sent in binary format. A binary response is adequate since framing is set by the requesting message, and because the master receiving logic knows exactly the number of bytes that should be received.

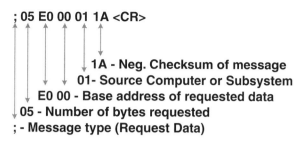

; 05 E0 00 01 1A <CR>

1A - Neg. Checksum of message
01- Source Computer or Subsystem
E0 00 - Base address of requested data
05 - Number of bytes requested
; - Message type (Request Data)

Figure 6.4. A blackboard request (read) data message

Blackboards and exposing parameters with symbolic addressing

In higher-level languages, the actual physical addresses of data are not constant. Instead, all data is linked to variable names at run time through a *symbol table*. For this reason, the previously described method of exposing variables is not easily accomplished.

One popular method for exposing symbolic variables through a protocol is to use arrays. In this type of scheme, all public variables of a class (like integers) are placed into an array and then simply referred to by the appropriate array index. These indexes are in turn given names. The resulting arrays can thus be treated as address blackboards for exposing the variables. In all but the simplest cases, these arrays will tend to be multidimensional. For example, instead of an integer called *Mode*, the system might have an array element, such as:

System_ Integers (Control, Mode)

The message protocol for requesting variables from these arrays will simply indicate the array to be accessed, the starting indexes, and the number of elements being requested. Note that each of these will be specified by an integer that had previously been equated to the array's index. For example, *Control* might be the dimension containing all of the control variables of the *System_Integers* array, and *Mode* might be the 3rd element of the control dimension.

In general, I have found this a very effective way of managing variables that are to be globally referenced, communicated and stored. For example, all the elements of the above array can be written to disk by simple loops that increment the indexes over their ranges, as opposed to writing long lists of individual variable names.

If these public variables are treated as individual class members instead of elements of arrays, then they will need to be read from and written to disk by their individual names, and there is no simple manner of requesting them in a message protocol. In large systems, this can lead to very long blocks of code that can easily become unmanageable.

Communications architectures

The communications architecture describes who initiates messages, and who passively receives them. There are two basic architectures: master-slave and peer-to-peer. By far, the most common and easy to implement architecture is the master-slave structure.

The master-slave architecture

The master-slave architecture operates like a classroom. The master initiates all communications and the pupils (slaves) respond. If a slave wishes to inform the master that it has data it would like to bring to the master's attention, it posts this fact in a flag that the master regularly tests. This is like a pupil raising a hand. When the master reads this flag it will make further inquiries as indicated.

A master-slave architecture is much simpler to implement than a peer-to-peer system because there is no danger of message collisions. Collisions occur when two agents both decide to transmit a message at the same moment.

Peer-to-peer architectures

In a peer-to-peer architecture, any agent on the link can send a message to any other agent. This architecture can be more responsive than the master-slave approach, but it is considerably more complicated to actually implement. Ethernet networks are peer-to-peer by nature.

In the master-slave approach, the master decides what the system is going to be doing, and it goes about sending the appropriate messages to the slaves. It will then monitor the status of each slave to determine if it can continue with the plan. If something goes wrong, the master finds this out at a predetermined decision point of its program.

With a peer-to-peer architecture, the managing process must be prepared to receive and act on a wide variety of messages at any time. This can greatly complicate the software design. For this reason, some systems are actually hybrids that allow only limited message initialization by the slaves.

There may be more than one communications link onboard the robot. For example, at Cybermotion we chose to use two links: a supervisory link and a control link. The supervisory link was mastered by the base station, and all of the computers on the robot responded as slaves. This link was used to program and monitor the robot. The control link was mastered by the mobile base and all of the other computers were slaves. The control link allowed the base to read sensors and direct subsystems. This architecture worked very effectively for us.

Wrappers, layers, and shells

The application protocol is almost never sent directly to the robot from the base station. Even if the message is sent by wire, it will be coded in a format such as RS-232. In reality it will travel over radio modems, telephone modems, the internet, or other means, and each of these will have its own protocol.

Some layers have to do with the way data is modulated onto carriers, while other layers are concerned with traffic handling, blocking, routing, and other matters. When the various protocols are part of a single master architecture (such as the factory MAP architecture), then these various protocols are usually called layers.

Most of these protocols are invisible to the user, but it is important to realize that the data stream may be broken into blocks for transmission. In fact, if you use a dial-up modem, it may use one of a variety of encoding protocols depending on line conditions. The timing of the application protocol decoders must be tolerant of this fact if the application protocol is to take advantage of all of the wondrous means of communications available today.

The last link from the base station to the robot will almost always be wireless. This can be infrared, FSK radio, or spread spectrum. The advent of inexpensive 802.11 Ethernet radios has opened up many possibilities for communications with mobile robots. The Ethernet protocol allows multiple data protocols to be exchanged with the robot at the same time but for different purposes.

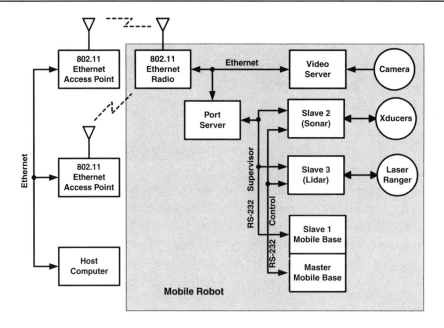

Figure 6.5. Simplified communications diagram of SR-3 Security Robot
(Courtesy of Cybermotion, Inc.)

For example, the robot's application protocol may be riding on an Ethernet link in a TCP format, while compressed audio and video are transmitted in parallel using their own streaming media formats as shown in Figure 6.5. In this case, all of the computers on the robot serve as slaves on the *supervisor link,* which is mastered by the *host* computer.

A second link is used for internal control of the robot in autonomous operations. This *control* link is hosted by the mobile base, and all other onboard computers are slaves. The control link does not extend outside of the robot. Both links use a superset of the protocol described in Figures 6.3 and 6.4. In this configuration, the host computer can communicate directly with the onboard slave computers for diagnostics and fine tuning.

Notice that the only physical communications channel from the host computer is an Ethernet link to multiple 802.11 radio Ethernet *access points.* Riding on the Ethernet protocol are multiple TCP connections for the video and data. At the vehicle, the port server changes the TCP stream by converting the data into a simple RS-232 stream.

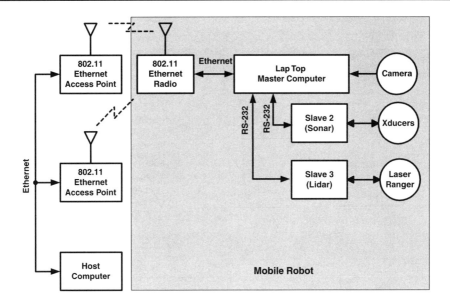

Figure 6.6. Communications structure using a laptop computer onboard

Figure 6.6 shows perhaps the most popular current communications configuration for robotics research. In this case, an onboard laptop computer serves as the main control for the robot, and only it communicates with sensor systems (usually by RS-232). This configuration has the advantage that the main onboard computer has a built-in screen and keyboard facilitating local diagnostics, as well as video capture. It has the disadvantage that the host cannot communicate with the sensor systems directly. It can also be a bit costly in production.

The direction that all of this is taking is toward systems like the elegantly simple structure shown in Figure 6.7. In this case, all of the sensor systems as well as the main computer use Ethernet communications. This configuration has the advantage that the base can communicate directly with the sensors. This assumes, however, that all the sensor systems are available with Ethernet interfaces. At present, this is not the case.

Another advantage in having the main onboard processor Ethernet equipped is that it can now communicate directly with other robots, door and gate controls, and elevator interfaces. There are some complications that occur as a result of all these intercommunications; for example, each robot must now behave politely in the connections it makes to shared resources. If this is not done, then these resources could become locked and unavailable to other clients and robots.

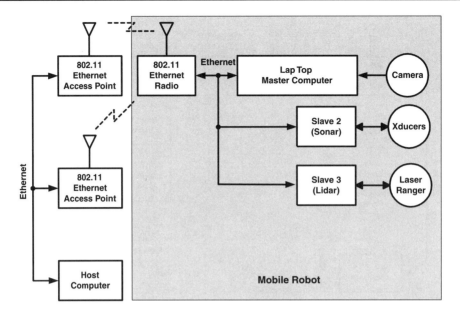

Figure 6.7. Communications structure based entirely on Ethernet

If one does not immediately appreciate the elegant simplicity that wireless Ethernet systems have brought to mobile robots, it is instructive to see how things were done only a few years ago. Figure 6.8 is the equivalent system as implemented for the SR-2 system fielded by Cybermotion before 1998. Since only low-speed radio modems were available, video had to be sent over separate channels. As robots moved through facilities, they would exceed the range of both systems.

Radio repeaters would relay the data channel to the robot, but at the expense of bandwidth. Video, however, could not be handled by repeaters and each receiver was connected to a switch. As the robot moved from location to location, the host computer would send commands to the switch to select the nearest video receiver. Actually, things were even messier than this when two-way voice communications were required. In these situations, yet another transceiver was added to the robot.

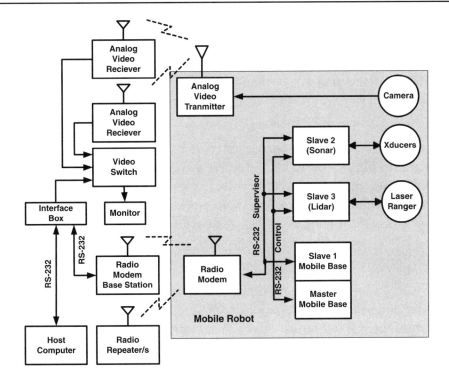

Figure 6.8. Communications diagram of SR-2 security robot (Circa 1997)
(Courtesy of Cybermotion, Inc.)

Drivers, OCXs and DLLs

When one thinks about the incredible complexity of detail that allows video, audio, and data to be exchanged seamlessly over an Ethernet link it would seem an impossibly complicated design process. Luckily, nothing could be further from the truth[5].

In the evolution of modern software, the issue of interfacing to a large number of systems with different proprietary applications protocols had to be addressed. At first, computer manufacturers tried to dictate protocol standards. These standards specified how each feature would be controlled. However, since manufacturers were always trying to offer more features than their competitors, attempts at standardizing such interfaces were like trying to nail jelly to a tree.

---

[5] When we switched our systems from analog video transmission and serial data radios to 802.11 and digital video, the process took less than two weeks.

To allow computer operating systems to be compatible with a wide range of peripherals, the burden was eventually placed on the hardware vendor to provide the interface. If the interface is at the hardware level (such as driving a printer through a serial port), the interface program is usually called a *driver*. Everyone who has ever installed a modem on their home PC has run into the process of loading the proper *driver*.

When the interface is at the software level, it is usually in the form of a DLL (dynamically linked library) or an OCX (active-X control object). A DLL is a library containing definitions and procedures that can be called from your application. You need to understand what each procedure does and in what sequence, and with what data they must be called. This process is usually straightforward, but exacting.

When an interface is provided as an OCX, it is in the form of a drop-in object that contains a completely encapsulated interface to the system. A good OCX will make you look like a genius[6]. The software programmer manipulates the OCX by setting its properties or calling its methods, and the OCX generates the application protocol messages to cause the desired result. All of the communications details are hidden from the programmer[7]. In many cases, the OCX will provide its own forms for the control of parameters, or the display of video, and so forth.

Porting your application protocol over the Ethernet is equally simple. In Visual Basic, you only need to drop a TCP host object into your project, and to tell it the address of the robot. If your link is to break out on the robot to a serial (RS-232) connection, this can be done with a small box called a port server. The port server IP address is entered into the host along with a port number. The port number usually defines the way the serial data will be handled. Port 6001 is typically used for straight-through connections.

Programming the transmission and reception of data with these "drop-in" hosts and clients is done very much as it is with serial communications. On the transmitting end, the data is simply loaded into the transmit buffer. On the receiving end, data arriving will generate *events*. These events are exactly like serial communications interrupt handlers, and if the status properties are okay, the data can be sent to your application protocol decoder.

---

[6] We used a digital video system from Indigo which had an excellent OCX.

[7] The proper solution to standardizing interfaces for the security robots mentioned earlier would have been to require each robot vendor to supply an OCX with certain properties and procedures, and not try to standardize the application layer protocol.

Improving communications efficiency

There are many ways to improve communications efficiency. In many cases, you will want to have more than one slave computer on a link, and there may be more than one link in your system (as in Figure 6.5).

Broadcasts

In some cases, a host computer will want to convey the same general information to more than one of the slaves. For example, the mobile base in Figure 6.5 may wish to *broadcast* the current position and heading estimate to all the sensor systems. In this way, slaves can process navigation information in ways not possible if they did not know the current position data.

The simplest way to perform a broadcast is to define a special message, and to have this message write into a special *broadcast bulletin board* in each of the slaves. This *bulletin board* can be a reserved block of memory, or an array of data. If communications becomes particularly busy, *broadcasts* can be transmitted at longer intervals, as long as they are sent reasonably often (every second or so).

The biggest problem with broadcast data such as position is that it is always out of date, and by varying amounts of time. In the case of position data, this can mean that any calculation made by the slave will be wrong because of this *latency*.

There are two ways to minimize this latency. The first method is to provide the slave with a predictive modeling program that runs at an interrupt rate much faster than the update rate of the broadcast. This modeler simply updates the position estimate in the *bulletin board* on the assumption that the robot is still accelerating (in each axis) at the same rate it was when the broadcast was received. In this way, the *bulletin board* values for the robot's position are continually updated for any task that might need them.

The second method of reducing broadcast data latency requires fewer system resources. This method involves time-stamping incoming broadcasts against a local time source, and provides a method or subroutine that is called in order to retrieve a current estimate of broadcast data. When the method is called, the program simply estimates the probable change for each axis since the data was received. This is similar to the first method, but the data is only refreshed as actually needed.

Flags

Flags are nothing more than variables (usually booleans, bytes, or integers) that can take on a limited number of values, each of which has a specific assigned meaning. Flags are a powerful tool in communications. Flags can be sent to slaves and they can be retrieved from slaves. One of the most common uses of such flags is to place them in a block of data that is regularly monitored by the host. The flag might tell the host that there is some other data, somewhere in the blackboard of the slave that it needs to look at. Flags can also request that the host perform emergency shutdown, or any number of other actions.

Two of the most common flags are the slave computer *mode* and *status*. By checking these integers or bytes, the host can tell if the slave is doing what it should be doing (*mode*), and if it is experiencing any problems doing it (*status*). Error numbers are another common example of flags.

To understand the power of flags, consider that the host computer might receive from a slave a simple byte representing one of 255 possible error codes. Although this code required only a single byte to be communicated, it may result in the operator receiving a whole paragraph of explanation about the nature of the error. A word of warning about flags is prudent at this point. If more than one of the conditions that a flag represents can be true at the same time, a numeric flag can only represent one of these conditions at a time. In these cases, the bits of a byte or integer can each serve as a flag, allowing any combination to be set at the same time. The other solution is to use individual flags for each condition, but this is very wasteful of bandwidth as an entire integer must be read to test each condition.

Templates

Although we have already indicated that we will try to group data contiguously in a slave's blackboard if we expect to request it at the same time, there will always be times when we want data from a range of different noncontiguous memory locations. When this is the case, the overhead for requesting a few bytes here and there can be very high.

Thus we need a *template*. A template is a script that the host loads into the memory of the slave in reserved locations. The script contains a list of starting addresses and the block size for each of these. The last of these pointers is followed by a zero-length block that serves to terminate the template.

The *template* is then requested by the host and the slave assembles the scattered scraps of data into a single message block. The efficiency of this messaging system is incredible, and it is easily implemented.

Post offices

In complex systems such as mobile robots, communications patterns shift with the nature of the task at hand. Furthermore, communications channels are *shared resources* that many tasks (clients) may need to use. In the case of a base station program, communications may be dictated by the number and type of forms that the operator has open. It is therefore intolerably inefficient for a particular thread to wait in a loop for the channel to become available, and then for the data to arrive.

In an efficient architecture, a task will request data *before* it absolutely needs it, and while it can still do useful work before the data arrives. To do *resource sharing* on a case by case basis rapidly results in unreadable code. Thus, we come to the general concept of a *resource manager*.

A *resource manager* for a communication channel is usually called a *post office*. A post office is a module (or better yet an object) that contains an array of pointers to messages and their associated flags, and destination addresses. The elements of this array are called *post office boxes*. At a minimum, a post office will have the following methods available for the clients to use:

1. Post a *request* message
2. Post a *send* message
3. Provide status of a posted message
4. Provide status of the communications resource.

As a task needs to communicate with other systems, it calls the *post office* and posts its message. The post office puts the message in the next available box, and then returns a message identification code that the client can use to request the status of its message in the future.

The reason that the post office does not simply return the *box number* is that boxes are reused as time goes on. If a client does not pick up its message information before the post office reuses the box, then the data in that box may have nothing to do with the client's request. In some schemes, the message identifier contains the *box number* in its lower bits to make it easy for the post office to locate the message when requested to do so by a client.

For a *send message*, a client provides the post office with a pointer to the message data, the size of the message, the remote processor (e.g., slave) to which the data is to be sent, and the destination address in that processor. The *post office* places all of this information into the next empty *box* and returns the message identification number to the client.

For a *request message*, a client provides the remote processor and address from which it wants data, the number of bytes it wants. It may also pass a vector where it would like the post office to save the received data.

When a message has been posted, the client will normally go about its business until it reaches the point of needing to know the status of its message. To determine this, the client calls the post office, and requests the status of the message by providing the message identification number. The post office will typically return one of the following status codes:

1. Unsent (Post office has not yet gotten around to the message.)

2. Pending (Message was sent but reply has not been received.)

3. Completed

4. Failed

Another useful method for a post office to provide is the status of the communications resource. This status usually indicates if the channel is connected, and the size of the backlog of messages it has to do. In this way, some clients can opt to slow down their requests of the channel when it becomes heavily loaded.

Advanced post offices sometimes support message priority. The priority of a message is usually limited to two or three levels, but can be greater. A higher priority message will get serviced sooner than a lower priority message.

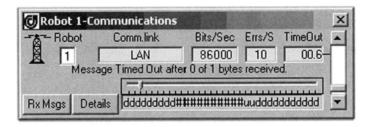

Figure 6.9. Post office monitor and diagnostics display
(Courtesy of Cybermotion, Inc.)

Message priority is handled in a simple way. The post office simply maintains a set of post office boxes for each level of priority. When the channel becomes available, the next message will be from the highest set of boxes that has a message posted. Thus, if there are no messages in the level 3 boxes, the post office will check the level 2 boxes, and so forth.

Many of the field problems in robotics are communications related, so it is important to be able to quickly determine the nature of the problem. The post office diagnostic display in Figure 6.9 shows all of the elements just discussed for a simple single-level post office. The bar at the bottom shows the state of all 32 of the boxes reserved for this robot's channel. The letter "d" indicates a message is done, while the letter "u" indicates it is unsent, etc. Clicking the "RxMsgs" button will show the most recent messages received that were rejected. The "Details" button gives statistics about which messages failed.

Although the concept of a post office may seem overly complex at first glance, it is in fact a very simple piece of code to write. The benefits far outweigh the costs.

Timing issues and error handling

Communications are seldom 100% dependable. This is particularly true if the messages must travel through more than one medium or by radio. Terrible things often happen to our messages once they are sent out into the cruel world. It is important that the various failure scenarios be considered in designing our communications structure.

Flashback...

One of the funniest communications-related problems I can remember was the case of the lazy robot. It was our first security robot and it was patrolling a relatively small, single-story laboratory. The main route was around the outside perimeter of the office, along an almost continuous window.

The robot would occasionally stop and refuse to accept any further orders. A security officer would be sent over in a patrol car, and would loop around the parking lot looking through the windows for the robot. By the time the robot was spotted, it would be running again!

The manager of the security force assured me that he was not at all surprised because this was exactly the way humans acted when they were given the same job! In fact, the problem was something called *multipath* interference.

Multipath interference occurs when the robot is at a point where radio waves are not being predominantly received directly from the transmitter, but rather as the result of reflections along several paths. If one or more paths are longer than the others by one-half wavelength, then the signals can cancel at the antenna even though the radio should be within easy range. When the patrol car approached, it added a new reflective path and communications was restored.[8]

Data integrity

The most obvious problem in communications is data integrity. A system can be made tolerant of communications outages, but bad messages that get through are much more likely to cause serious problems. There are many forms of error checking and error correction, ranging from simple checksums to complex self-correcting protocols.

It is useful to remember that if an error has a one in a million chance of going undetected, then a simple low-speed serial link could easily produce more than one undetected error per day! Luckily, our application protocol is likely to be carried by protocols that have excellent error detection. Even so, it is useful to discuss error checking briefly.

The two most popular error checking techniques are the checksum and the CRC (cyclical redundancy check). Checksums are calculated by merely adding or (more commonly) subtracting the data from an accumulator, and then sending the low byte or word of the result at the end of the data stream. The problem occurs when two errors in a message cancel each other in the sum, leaving both errors undetected.

A CRC check is calculated by presetting a bit pattern into a shift register. An exclusive-or is then performed between incoming data and data from logic connected to taps of the shift register. The result of this exclusive-or is then fed into the input of the shift register. The term *cyclical* comes from this feedback of data around the shift register. The function may be calculated by hardware or software.

When all the data has been fed into the CRC loop, the value of the shift register is sent as the check. Various standards use different register lengths, presets, and taps.

---

[8] The multipath problem disappeared in data communications with the advent of spread-spectrum radios, because these systems operate over many wavelengths. The problem is still experienced with analog video transmission systems, causing video dropout and flashing as the robot moves. All of these problems go away when communications is combined with video on 802.11 spread-spectrum Ethernet systems as shown in Figure 6.5.

It is mathematically demonstrable[9] that this form of checking is much less likely to be subject to error canceling.

The ratio of non-data to actual data in a message protocol is called the *overhead*. Generally, longer checks are more reliable, but tend to increase the overhead. Error correcting codes take many more bytes than a simple 16-bit CRC, which in turn is better than a checksum. The type and extent of error checking should match the nature of the medium. There is no sense in increasing message lengths 10% to correct errors that happen only 1% of the time. In such cases it is much easier to retransmit the message.

Temporal integrity

A more common and less anticipated problem is that of temporal integrity. This problem can take on many forms, but the most common is when we need data that represents a snapshot in time. To understand how important this can become, consider that we request the position of the robot.

First, consider the case where the X and Y position are constantly being updated by dead reckoning as the result of interrupts from an encoder on the drive system. Servicing the interrupts from this encoder cannot be delayed without the danger of missing interrupts, so it has been given the highest interrupt priority. This is in fact usually the case.

Now assume the following scenario. The hexadecimal value of the X position is represented by the 32-bit value consisting of two words, 0000h and FFFFh, at the moment our position request is received. The communications program begins sending the requested value by sending the high byte of 0000h first. At that moment an encoder interrupt breaks in and increments the X position by one, to 0001h and 0000h. After this interruption, our communications program resumes and sends the low word as 0000h. We receive a valid message indicating that the X position is 0000h, 0000h, an error of 32,768!

We cannot escape this problem by simply making the communications interrupt have a higher priority than the encoder interrupt. If we do this, we may interrupt the dead reckoning calculation while it is in a roll over or roll under condition, resulting in the same type of error as just discussed.

---

[9] The author makes no claim to be able to demonstrate this fact, but believes those who claim they can!

For this and other reasons, it is imperative that the communications task copy all of the requested data into a buffer before beginning transmission. This transfer can take place with interrupts disabled, or through the use of a data transfer instruction that cannot be interrupted.

There are even more subtle forms of errors.

Flashback...

I am reminded of one of the most elusive bugs I ever experienced. The problem showed up at very rare times, only in a few installations, always along the same paths, and in areas with marginal radio communication. The robot would receive a new program, begin executing it, and suddenly halt and announce an "Event of Path" error. This error meant that the robot had been instructed to perform an action as it drove over a point on a path, but that it did not believe the point was on the path! More rarely, the robot would suddenly double back to the previous node and then turn back around and continue on. It was very strange behavior for a major appliance!

This bug happened so rarely that it was at first dismissed as an observer-related problem, then as a path-programming problem. Finally, a place was found where the problem would occur fairly regularly (one time in 100) and we continued to exercise the robot until we were able to halt the robot at the problem and determine the cause. This is what was happening:

When the robot finished a job, it would halt. At that point, it would be sent a new program and then it would be told to begin execution of the program at step one. The program was loaded in blocks, but no block would be sent until the block before it had been successfully transmitted. Once the whole program was successfully loaded, the instruction pointer would be set to the first instruction and the mode would be set to automatic. As a final check, the program itself contained a 16-bit checksum to assure that it had been put into memory without modification.

The problem, it turned out, was caused when the robot was sent the message setting its program pointer and mode. If this message was not confirmed within a certain time, it was retransmitted. It had been *assumed* that a message timeout would be the result of this message not being received by the robot. In that case, the code worked fine. The real problem came when the robot *did* receive the message, but its reply saying that it had received the message did not get back to the host. The host would then wait a few seconds and retransmit the reply, causing the robot to jump back to the beginning of its program like a record jumping a track[10]. Thus, it would begin the program over even though it might now be 50 feet from where that program was supposed to start. If it had crossed through a node before the second message arrived, the robot would throw an error. If it had not reached a node, then the second transmission would not cause a problem because the program pointer was still at the first step anyway.

The result was that the problem only occurred in areas of poor communications where the distance from the first node to the second node was relatively short. The reason it was so hard to detect was because it occurred so rarely and because of the *assumption* that the message did not go through if a reply was not received. This was a classic *assumption* bug.[11]

Other issues

There are other issues that arise in some situations. One of these occurs when communication is routed over a network that has excessive message persistence. If the protocol handler or post office has given up on a message due to a time-out, but one of the networks along the way is still repeating the message, then a return message may eventually arrive that is badly out-of-date. Worse yet, we may believe that it is the reply to a new message, causing great confusion. For this reason, if possible, it is best to set the message persistence of such systems to a minimum value.

Another safeguard involves tagging messages with message numbers so that it can be determined for sure whether a reply belongs to a particular message. If a messaging system is full duplex, then messages are being sent without waiting for replies, and this type of tagging is essential.

Books have been written about single aspects of the issues discussed here. The important thing is to recognize the value of a good communications system, and to anticipate as many problem scenarios as possible in structuring it. Once the structure has been selected, you can research specific areas to the depth required. It is also crucial not to fall into the trap of sacrificing bandwidth for development convenience. It may seem that a system has more than enough bandwidth to allow such waste, but as it grows this will almost universally prove to be untrue.

---

[10] For the reader who is not eligible for senior discounts, I should explain that records were an ancient means of storing and playing back music. They were plastic disks imprinted with grooves whose bumpy sides crudely represented a series of sounds.

[11] See Chapter 17.

Section 2:
Basic Navigation

Basic Navigation Philosophies

Before we plunge into the subject of navigation, it is helpful to put the whole topic into perspective. Thousands of papers have been written about robot navigation, and many very compelling concepts have been developed. When one looks at the actual robotic products that have been produced, however, most have used much simpler approaches. Two schools of thought about robot navigation have emerged: the academic school and the industrial school.

The academic school of thought

There is a recurring theme in many academic papers on robot navigation that goes like this: "Before robots can be widely accepted, they must be capable of learning their environments and programming themselves without assistance." So consistently is this line repeated, that one is prone to believe it must have been chiseled on the back of the tablets that Moses brought down from the mountain. Whether it will turn out to be a fact, however, is yet to be determined. There is evidence on both sides of this argument.

In support of the academic argument, it is true that most potential customers (other than the academic and hobby markets) do not want to program their robots. These customers, quite rightly, do not want to be distracted from the primary mission focus of their business or department. On the other hand, they do not care whether a robot is programmed or not, as long as it is installed and does the job they expect. The yardstick for such customers is one of return on investment (ROI). If programming takes a high degree of expertise on site, then this will add to the cost of the initial installation and certainly reduce the ROI. It is also important that the system be readily changed when the environment changes. If changes add to operational costs, then the ROI is reduced again.

To those in the academic camp, the robot's sensor systems should recognize the entire environment, not just a feature here and there. They envision the robot as perceiving the environment in very much the same way humans do. Indeed, this is a wonderful goal, and some impressive progress has been made in video scene analysis. Figure 7.1, however, shows an example of an environment that severely challenges this goal.

Figure 7.1. SR-3 Security robot navigating from lidar (circa 1999)
(Courtesy of Cybermotion, Inc.)

In this example, the robot is using lidar to navigate from pallets of products. The robot patrols evenings, and is idle during the day. Thus, on successive evenings the presence and placement of pallets may change radically. In this case, the robot has been programmed to look only for pallets that lay near the boundary of the aisle. The laser scans parallel to the ground at waist height. In the event that there are not many pallets present, the robot's programming also tells it about reflectors on a wall to its right, and on vertical roof supports on the left (just visible at the top of the image).

The challenge for a hypothetical self-teaching system is that almost the entire environment changes from night to night. If such an ideal robot had a video processor on board, and was smart enough, then it might notice that the painted stripe does not change from night to night and learn to use it as a navigation feature. For now, however, the safe and easy way to solve the problem is for someone to tell the robot which features to use ahead of time.

There are many ways to solve any one problem, but the trick is to create systems that economically solve the problem reliably for multiple environments with the least expensive, lowest power, and most reliable sensor systems possible.

The industrial school of thought

The industrial school of thought is diametrically opposed to the academic school. This view is often verbalized as KISS (keep it simple, stupid). Many in this school believe that if a guidance system works well by following a physical path, then why bother with all that fancy navigation nonsense. There are, in fact, applications where these old technologies have sufficed for decades.

People of the industrial school are business people who see themselves as providing solutions to customer problems. They understand that the cost of development is very high, and that the return for such investment is often insufficient. Unfortunately, for those who prescribe too rigidly to this school, recent techno-history has clearly shown that sitting still is an ever more risky survival strategy. The inherent rigidity of physical path systems makes them potentially vulnerable to more flexible and capable autonomous systems.

Similarly, in the case of emergency response and bomb disposal robots, there is likely to be an evolution toward systems that make the remote operator's job easier and the robot safer. The simple addition of collision avoidance sensors and reflexive behaviors could make these robots safer and easier to drive. Such improvements would also reduce the required operator training costs, allowing more cross training with other jobs. There is absolutely no technical barrier to this approach, but most of the manufacturers in this field are still producing robots with little or no automatic capability.

In the end, the competition between these schools of thought will be settled in the marketplace on an application-by-application basis. The result overall will most probably be systems that are increasingly easier to install and more flexible, while becoming less and less expensive. It is unlikely that the whole market will wait for the perfect robot, and then burst upon it in wild enthusiasm.

Area coverage robots

Area coverage robots for applications such as lawn mowing and vacuum cleaning have already been marketed for years with various degrees of success. These devices require no programming, and use a variety of strategies to cover the area they are placed into.

At the low end, these robots have incorporated strategies that do not represent true navigation, but rather movement patterns that should eventually cover the area. For example, lawn-mowing robots have been developed that randomly cross an area until they detect a boundary marker and then turn away to cross the area again. While most of the area is covered quickly, some areas may never be covered. The method is extremely inefficient.

The Roomba™ vacuuming robot produced by iRobot uses a much more effective area-covering strategy to cover a contiguous area, and has a rudimentary sense of position. For the job of cleaning a room in one's home, this strategy is sufficient, and the price is low enough that the robot does not need to be more independent to be cost effective. The only programming required is to set the room size and possibly install an infrared beam called a "virtual wall" to prevent a robot from moving into undesirable areas.

Figure 7.2. Roomba™ area vacuum robot

(Photo courtesy of iRobot, Inc.)

At the high end, cleaning robots have used sensor-based local navigation that provides efficient area coverage, but only within a contiguous and bounded area. If these robots are not to enter a certain area, then physical barriers such as roadway cones are used to accomplish this.

In none of these cases, however, has the robot been capable of understanding its global position and going from one area to another to do things in a specific manner. To do this, the robot must by very definition be programmed in some way.

Virtual path following vs. goal seeking

There are two basic philosophies about how a robot should find its way from one place to another: *virtual path following* and *goal seeking*. Goal seeking has traditionally been the favorite method of academia. In goal seeking the robot knows where it wants to go in terms of coordinates and it plans the best route to achieve the goal. In doing so, it may or may not be provided in advance with a map of known obstacles. In any case, as it encounters new obstacles it will build a map of them.

Virtual path following

Virtual path following is essentially like physical path following except that the path is defined in the programs of the robot and not by painting a stripe on the floor. A virtual path-following robot will typically have a path program defined either graphically or as a text program or both.

Cybermotion robots, for example, execute pseudocode programs that require the transmission of very little data. These programs are generated using a graphical environment called PathCAD (Figure 7.3). In the PathCAD environment, graphical objects that are dropped onto the drawing have properties that are carried invisibly with the drawing. These objects include *nodes* with names such as V1, V2, and E3 on Figure 7.3.

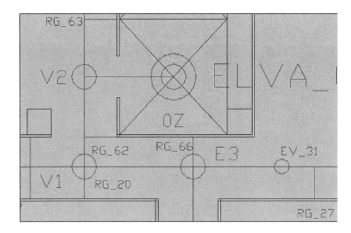

Figure 7.3. PathCAD programming environment for virtual paths
(Courtesy of Cybermotion, Inc.)

Once *nodes* and the paths connecting them have been drawn, a graphical aid is used to generate text "path actions." These *actions* are then compiled into the path programs that the robot receives. We will discuss this process more in later sections, but a few objects are worth noting.

Besides the *nodes*, objects of interest in the example shown include the event EV_31 and the range definitions RG_20, RG_62, and RG_66. Events tell a robot to do something at a point along a path. Range lines are used in the navigation process. The most complex object is the *elevator node* in the center of the drawing. This node references all of the programs required to operate the elevator and to take the robot from one floor to another.

Virtual paths are more than a halfway step between physical paths and goal seeking. Since the robot knows where it is, it can also have behaviors for circumnavigation that are very similar to goal seeking.

Goal seeking

The method of finding the shortest route through obstacles has traditionally been based on one of several models, including "growing obstacles" and "force fields." Figure 7.4 demonstrates the concept of growing obstacles.

The precept is simply that the size of all obstacles is swollen by the clearance required for the robot. Once this is accomplished, one has only to assure that there is a finite space between the swollen obstacles to plan a path between them. If the robot is oblong, then one has to decide which dimension to use to grow the obstacles. If the width is less than the length (the usual case), then it may be tempting to use the width as the growth dimension. If this is done, however, then the robot may not be able to turn around in some tighter places.

Route planning by growing obstacles

The *obstacle growing* method is essentially binary, in that the planner tries to make the path go as directly toward the goal as possible, and in so doing pushes the path to the very edge of the swollen obstacles. The process yields the shortest, but not the safest or smoothest path.

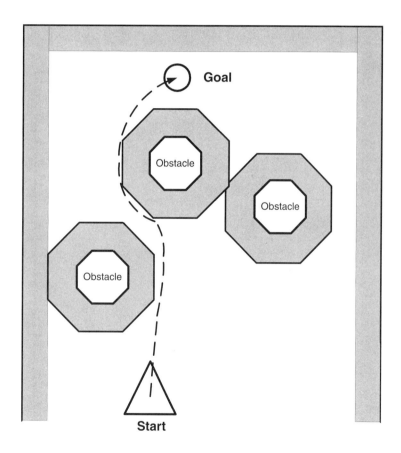

Figure 7.4. Growing obstacles for path planning

Route planning by force fields

Another popular approach is based on calculating imaginary repelling fields emanating from obstacles as shown in Figure 7.5. The fields can diminish linearly with distance or geometrically. The process typically yields a path that stays as far away from obstacles as possible. Limits are normally placed on the range of the field effects, so that calculations do not need to include distant objects.

In order to visualize this process, it helps to imagine the resulting map as a landscape where the obstacles are ridges or peaks, and the most likely routes follow the valleys (Figure 7.6). The floor of a valley occurs where two or more force fields cancel each other and the height of a valley floor is equal to the magnitude of the canceling fields. The path planner is allowed to route the robot through a valley as long as its height is below a minimum. This minimum corresponds to the clearance of the vehicle.

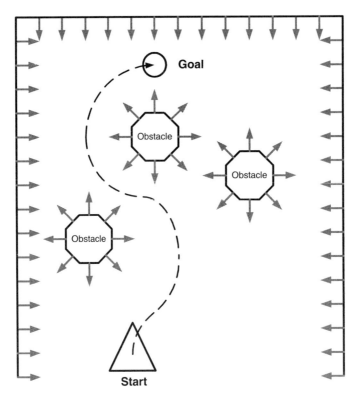

Figure 7.5. Force fields for path planning

Disadvantages of goal seeking

Goal seeking is elegant and interesting, but it assumes that the only reason for choosing a route is to avoid obstacles. In most applications, there are other reasons for choosing a particular route as well. For example, a security robot's route is planned to assure that the robot performs surveillance on all important areas. In other cases, there may be traffic in the area that requires rules of the road to be followed.

As mentioned earlier, the major disadvantage of goal seeking is that it takes routing decisions out of the hands of the system programmer. Therefore, if a robot is to patrol a building in a specific manner using goal seeking, then it must be given a sequence of goals that force it to cover the areas desired.

A practical starting point and "a priori" knowledge

Returning to the question of whether our robot should be self-teaching or preprogrammed, we must again ask the following question: "If our robot were to map its environment without help, and be totally independent and self-programming, then how would it know how to call and board an elevator or open a door?" Even if it could be made smart and dexterous enough to perform these tasks, how would we communicate what we wanted it to do?

So, which of these concepts and techniques represents the best approach? In the end, the most practical method of programming is one that merges many of the qualities of both the academic and industrial camps, the virtual path-following techniques and the goal-seeking techniques. Like so many decisions in robotics, there is a natural answer and no matter which camp you start in you will be drawn toward the same consensus.

Our approach will therefore be one that lies somewhere between the extremes just discussed. Since wonderful maps and floor plans now exist in digital form for almost every public place, we will start with these drawings as a basis for the programming approach. The data on these maps and drawings is far more global and detailed than our robot will ever be able to produce on its own. If we wish our robot to participate in the mapping process, then let's give it what we know and let it build its maps on top of this information.

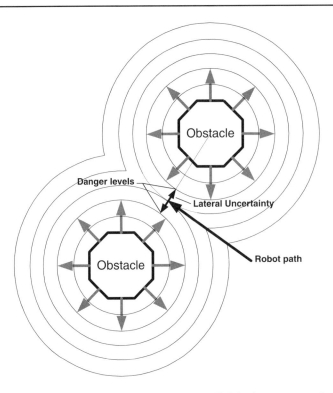

Figure 7.6. Topological map of force fields between obstacles

We will then begin by telling our robot where we would like it to go, and what we would like it to do. This is no different than what we might do for a new human worker, except that our means of communicating this information will be a bit different.

We will also help the robot with the selection of features it should use to navigate. We know which objects in its environment are permanent and which are not, so a little help from us can greatly improve the robot's performance. If we find a method that allows the robot to use its sensors to add navigation features of its own, our architecture should make it possible to add this capability at a later time.

CHAPTER 8

Live Reckoning

It is ironic that the most central precept of autonomous navigation is often viewed disparagingly, especially by laypersons. Perhaps this is because the term *dead reckoning* was adopted to describe this precept, and most people have a preformed view of what this entails. The very name *dead reckoning* seems to imply doomed guesswork. At best, the term conjures up images of Charles Lindbergh flying across the Atlantic Ocean with little more than a compass heading for navigation.

If we announce that our robot has excellent *dead reckoning,* many in our audience will think this means that our robot *has no true navigation at all!* At the very least, they will tend to think that our robot, like Lindbergh, travels long distances between updates of its position. For the purposes of this discussion, we will surrender to this view, and use the new term *live reckoning* to describe what we want to accomplish. To under-stand *live reckoning,* we should first discuss *dead reckoning* a bit.

The truth is that dead reckoning was invented by nature long before man coined the term, and this is just the first of many of nature's techniques that will be useful to us. One of the Zen aspects of mobile robot design is the fact that we will continually come to these same concepts whether we attempt to copy nature or not!

In mammals, dead reckoning is accomplished primarily by signals from the inner ear. We may think we find our way across a room by visual navigation, but the fact is that without our inner ear we couldn't even stand up and walk. Other senses, such as the feeling of pressure on the soles of our feet, play into our navigation next, and finally our vision is used to correct these estimates.

Inertial navigation uses accelerometers and gyroscopes as inputs to its calculations, but is in the end a form of dead reckoning. On the other hand, if the position and orientation estimate of an inertial navigation system is being regularly updated with beacon or GPS position fixes, then it becomes *live reckoning*. The distinction between the two is therefore largely a matter of frequency, but it is an important distinction.

To avoid the confusion we will avoid the term dead reckoning. The process of registering relative motion on a continual basis will be called *odometry* instead.

Relative vs. absolute

At this point, we must differentiate between *relative* navigation information and *absolute* information. Odometry is relative information that tells us how much we have moved, but not where we are. If we know where we started, then we can add the relative movement to the starting position and obtain an estimate of our current position.

Odometry gives us very fast feedback, but because of accumulated errors, it is subject to drift. Other navigation techniques will be needed to correct the odometry estimate periodically. Some of these may correct all axes, and some may correct one or two.

These principles have been used for much of recorded history. Late in the Battle of Britain in World War II, the Luftwaffe decided to switch from day bombing to night bombing to reduce their rather severe losses from Royal Air Force fighters. Since England was under a strict blackout, there were no natural landmarks from which German pilots could correct their dead reckoning navigation, especially on cloudy or moonless nights.

Thus, the Luftwaffe developed a ground-based beam system code named "Crooked Leg." The bombers "rode" a primary pair of beams that provided an indication showing whether they were straying to the left or right of their intended path to the target. On the left the pilot would hear Morse code dots, and on the right dashes. In the middle, these signals would combine to provide a continuous tone. A second beam pair intersected the first just before the target at the IP (initial point), telling the crew to prepare to drop their bombs[1].

---

[1] The British quickly detected the beam system and added a signal of their own to cause the indicated path to be deflected off of the target. The history of WWII is rife with fascinating examples of technologies and counter measures being developed in rapid succession.

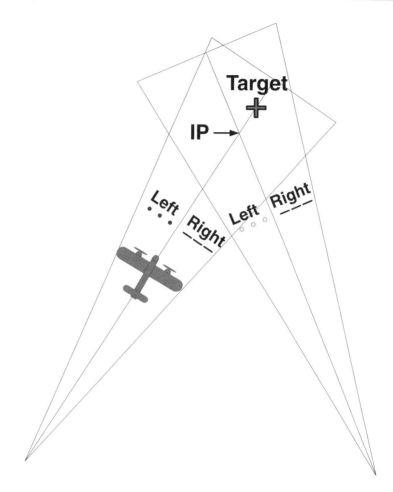

Figure 8.1. The "Crooked Leg" navigation system

In navigation terms, the first beams gave the pilot absolute information about the aircraft's *lateral* position with respect to its intended path, but told nothing about the *longitudinal* position along the path. The longitudinal position was estimated en route from air speed and wind reports. The second beams provided the *longitudinal correction* or *longitudinal fix*. These basic concepts are the same as those of mobile robot navigation, which in turn are the same as those used by animals for navigation.

One of the very few absolute navigational references available is the GPS (Global Positioning System). This system provides very accurate absolute position information by triangulating satellite signals. Best of all, it requires only an inexpensive GPS receiver.

At first glance, this capability would seem to offer a total navigation solution, but this is not the case. GPS signals can be blocked by overhead structures, severe weather, and even heavy vegetation. GPS is of no use at all in most indoor environments. If a robot is to depend totally on GPS, it will not be capable of operating in many environments and situations. GPS is an immensely powerful tool for outdoor navigation, but even this technology must be integrated into a navigational architecture with other techniques.

Understanding why good dead reckoning is crucial

Live reckoning is the heart of any good navigation system, providing continuous position updates between *fixes*. This position estimate is then used as a basis from which to evaluate incoming navigation data. If a platform provides good odometry, then it will have even better *live reckoning*.

If our robot is navigating from GPS, and suddenly passes under a highway overpass, good odometry can keep it rolling until the GPS signal is restored or the vehicle receives some other kind of navigational data. All else being equal, the better the odometry, the more robust the navigation will be. In indoor navigation, robots tend to depend on odometry even more significantly as they move from one partial *fix* to another.

Platform inherent odometry

The better a vehicle's odometry, the better it will be able to navigate. The quality of a mobile platform's inherent odometry is dependent on the drive system configuration. The most critical axis is steering and the sense of direction because errors in this parameter will accumulate geometrically as the vehicle drives.

The less coupling a platform has between steering and driving, the more predictable will be its response to a steering command.

Vehicles using Synchro-drive have independent motors for driving and steering. The drive motor causes all of the wheels to turn at the same rate, while the steering motor causes all the wheels to perpetually steer in the same direction. Since driving and steering are decoupled, the vehicle responds precisely to steering commands, even on very slippery surfaces. A well-aligned Synchro-drive platform will seldom require additional heading sensors.

The worst configurations for odometry are skid-steered vehicles that have a high degree of coupling between driving and steering, and also depend on slippage between points of contact and the ground

A tank, for example, has no steering as such, but depends on driving its two treads at different speeds to accomplish steering. If one tread has better purchase than the other, the intended amount of turning may not occur. The same is true for wheeled vehicles that use skid steering and differential drive[2].

A vehicle using Ackerman steering (e.g., a common passenger car) has a steering mechanism that is independent of the drive mechanism, but depends on the vehicle moving forward (or backward) while the steering is deflected before the steering has an affect on its heading. If the car's drive wheels slip, and it does not move forward the amount indicated by its odometry, then a heading error will occur. For this reason, Ackerman systems are somewhere between Synchro-drive vehicles and skid steered vehicles in the quality of their *inherent odometry*.

The choice of a drive system may have many other considerations besides its inherent odometry quality. If the system does not have good inherent odometry, then it is essential to supplement the odometry with the appropriate sensors. One of the most common sensors for this purpose is the gyroscope. Optical gyroscopes are now available that can provide the vehicle with an excellent sense of direction. Unfortunately, the cost of an optical gyroscope is usually only reasonable in higher-end applications.

Picking coordinate systems

A robot will normally use two coordinate systems, one for global reference and one for local reference. The global reference system is concerned with where the robot is in its environment and it has no inherent center. For this reason, global systems are almost always Cartesian.

The local reference system, on the other hand, is used to integrate data from sensors into the robot's sense of things. This process involves vector math, and is inherently centered on the robot. Therefore, the natural coordinate system for this purpose is a polar system.

---

[2] Skid steering uses two or more powered but non-steered wheels on each side of the vehicle. Differential drive uses two powered wheels on either side of the platform with the other wheels being casters.

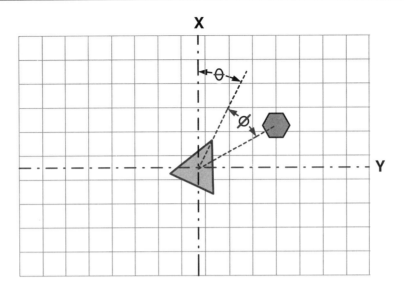

Figure 8.2. Absolute Cartesian and relative polar coordinate systems

As an example, consider the radar on a ship. The center of the screen is in the middle of the ship, and its display uses a polar coordinate system. On the other hand, the navigator calculates and reports the ship's position in latitude and longitude, a spherical system that looks like a Cartesian coordinate system over most of the globe (except near the poles). Maps always use a Cartesian grid system for reference.

Similarly, when we observe things, we think of their position relative to our position and orientation, but when we want to convey this to others, we describe their positions relative to things having global coordinates. As in the choice of live reckoning for the first level of navigation, the choices of these coordinate systems are virtually preordained.

Tick calculations

At its lowest level, odometry is almost always accomplished by encoders on the drive and/or steering systems. Occasionally, platforms will use encoders on nondriven idler wheels that contact the ground below the vehicle. These configurations are usually an attempt to eliminate the odometry errors resulting from drive-steering coupling. Unfortunately, these configurations tend to be overly complex and unreliable, and have not generally found widespread usage.

An encoder *tick* is a change of a drive encoder by the amount of its smallest increment. A position change smaller than a *tick* will not be detected, and the motion associated with a tick will normally be approximated as a straight *tick vector* to be added to the platform's position estimate. The higher the resolution of an encoder, the smaller the errors produced by these approximations will be.

Encoders are usually binary in nature, so their resolutions can be any power of two. Typical resolutions range from 32 to 4096. Signals associated with the act of driving generally cause a *tick interrupt* to the navigation computer. Thus, for drive encoders, the important issue is the amount of travel represented by a single *tick*. If the *tick distance* is made too large, the odometry calculation will be inaccurate because it will be incapable of tracking fluctuations in the vehicle's course during the tick interval. On the other hand, if the resolution is needlessly small, the encoder will cause an excessive overhead for the computer it is interrupting. As a rule of thumb, the *tick distance* should be about one thousandth of the distance between the wheels or treads of the vehicle.

For those systems with steering encoders, the angle of the *tick vector* will be truncated (not rounded) to the resolution of the encoder. It is highly desirable that the steering encoder be mounted as close to the steered wheel(s) as possible, so that it is not affected by the mechanical backlash of the steering mechanism. The resolution of the steering encoder should be about a factor of three smaller than the backlash between it and the actual wheels. Typical resolutions will range from 1024 and up.

Figure 8.3 demonstrates many of the issues just discussed. In this case, the right tread has remained stationary while the tread on the left has driven one tick of distance *d* as represented by the large arrow.

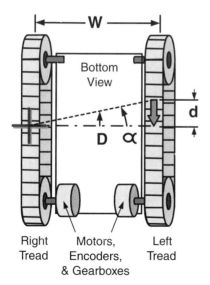

Figure 8.3. Odometry tick of skid-steered platform (seen from below)

First, we must determine the relative movement of the platform that probably resulted from the drive *tick*. This action has two effects; it both drives and turns the platform. We will assume that traction is evenly distributed across both treads, so the platform pivots about the center of the stationary (right) tread. The center of the platform moves forward by the distance D, which is equal to half the motion of the driven tread:

$$D = d / 2$$

Note that D is actually an arc, but if the tick distance is small enough it can be assumed to be a straight forward move. By approximating all moves as minute straight vectors at a single heading, we are able to track any complex maneuver.

Because of this action, the platform steers clockwise by the angle α (remember we are looking from below). This angle is simply the arcsine of the tick distance that the right tread moved, divided by W, the distance between the treads. If the traction across the treads is not even, then the platform may not pivot as shown, further demonstrating the difficulty with skid steered odometry.

$$\alpha = + \text{Arcsine} (d/W) \quad \text{for the left tread}$$

$$\alpha = - \text{Arcsine} (d/W) \quad \text{for the right tread}$$

If this action by the left tread is followed by a tick of forward motion by the right tread, then the angular changes will cancel and the platform will have moved straight forward by the distance of a tread tick.

If instead, the forward movement of the left tread is followed by a reverse move of the right tread, then the angular components will add while the linear motion elements will cancel. In other words, the platform will turn on center (hopefully).

The next step is to add this relative motion to the vehicle's current position and heading estimate. The relative platform motion is a vector of D magnitude at the current heading (θ) of the platform. This vector must be summed with the position of the vehicle to produce the new Cartesian position estimate. The angle α is then added to the heading of the platform to obtain the new heading.

For a platform operating on a flat surface, the vector calculation is:

$$x = x + (D * \text{sine}(\theta))$$

$$y = y + (D * \text{cosine}(\theta))$$

$$\theta = \theta + \alpha$$

Where x and y are the Cartesian coordinates of the platform position and θ is its heading.

Note that the direction of a positive angle in navigation is usually clockwise, the opposite convention of that used in geometry. It is not important which system is used, or along which axis the zero heading is assumed, as long as these are used consistently. I prefer the "map" system, with zero heading being up the positive y-axis, and positive angles being clockwise.

If the platform is operating over uneven terrain, it may be necessary to take into account the *pitch* angle of the platform. In this case, the vector distance D would be multiplied by the cosine of the pitch ρ. The y-axis (altitude) position estimate of the platform would be changed by the sine of the pitch multiplied by the distance D.

$$x = x + (D * \text{sine}(\theta) * \text{cosine}(\rho))$$

$$y = y + (D * \text{cosine}(\theta) * \text{cosine}(\rho))$$

$$z = z + (D * \text{sine}(\rho))$$

The above calculations are valid for most skid-steered and differential drive platforms. With Synchro-drive vehicles, the process is even simpler, as the drive system is inherently a polar coordinate system by nature. The angle of the steering encoder plus an offset angle yields the heading of the vehicle directly. Since these calculations may need to be executed many thousands of times per second, it is important to consider a number of factors.

The tick interrupt

If the interrupt is not processed quickly, it may *overrun*, or the next interrupt may not get processed. Therefore, the *tick* interrupt(s) should normally be the highest priority interrupts on the processor. This interrupt should never be masked for more than a few cycles at a time, and it should itself be uninterruptible. It is, however, important that this interrupt never occur when another process is accessing the position or heading estimates, as this may cause erroneous values to be used. Processes that must access this data should do so in a block move if possible.

If an interrupt is not available, it may be tempting to *poll* the encoder. This is a really bad idea, as it will consume enormous resources from the CPU, but it can be done if such resources are available. A better solution may be to place a small, dedicated micro-processor between the encoders and the main CPU, and to program it to perform the calculations. Processors are available for a few dollars that can handle this process.

Speeding up calculations

The first thing we need to know is how fast our processor can perform these calculations. If this is not known explicitly, then we can devise a test that will exercise the CPU and determine the amount required to perform, say, one million such calculations. Next, determine the highest rate that the interrupts could occur, and make sure that the time is well more than an order of magnitude shorter than the shortest period for the interrupt.

The two most common ways to increase the speed of such calculations are by adding a coprocessor, or by using lookup tables for the geometric functions. In the calculations above, notice that the arcsine function is always called with the same value (d/W). It is silly to ask our processor to do this arcsine calculation over and over. The obvious solution is to pre-calculate this constant, and then just manipulate its sign.

Live reckoning interaction with other processes

We achieve *live reckoning* when odometry begins interacting tightly with the other processes of navigation. The position estimates produced are used by sensor systems to orient the data that they are processing. These sensor systems may in turn correct one or more of the baseline position and heading estimates. Since the odometry process is one of adding relative movements to these estimates, the algorithm must pay no attention to the fact that the estimates have been changed by someone else.

The position and heading estimates are also used by higher-level systems for such purposes as determining if a destination has been reached, or showing a remote operator the position of the robot. Other, less obvious uses of this data will be discussed in the chapters that follow.

The uncertainty estimates for each axis of a robot are almost as important as its position estimates for these axes. In the coming chapters, we will explore the many ways that the uncertainty model will interplay with navigation and behavior.

CHAPTER 9

The Best Laid Plans
of Mice and Machines

Having trampled our way through all of the requisite background subjects, we arrive at last at the fun part. We must now structure a software environment that enables our robot or vehicle to move purposefully and safely through its environment, using data that is often incomplete and even contradictory. This structure must also be flexible enough to undergo evolution as sensors and algorithms evolve.

The answer is to create an almost schizophrenic process, in which one half is the optimistic planner and executer, and the other half is the skeptical critic and back-seat driver. Navigation in autonomous robots is actually two almost completely separate processes that take place at the same time. The first process assumes that the position estimates are correct, and executes a plan. The second process monitors the progress and sensor systems, and corrects the odometry estimates as appropriate. First, let's consider planning.

The trick to the planner part is that the planning process must be capable of having its preconditions changed on a continual basis without becoming disoriented. This means that the process is *dynamic*, keeping little static information about its previous plans. If it had been planning to converge on a path from the left, and the planner suddenly finds it is mysteriously already on the right side of the path, it must simply adapt its approach without missing a beat.

There are two immutable laws we must accept:

1. A robot is never exactly where its position estimate says it is.

2. A robot's position estimate is never exactly where it wants to be.

The first law is due to the fact that navigation is never an exact science, and the readings from sensors will always contain some error. Furthermore, the robot may experience stretches on which it has no active navigation and is depending entirely on odometry.

The second law is due to the fact that it is impossible to move a mass to an exact position[1]. We must accept the fact that our performance will always contain some error, and set our goal for "good enough."

Path planning and execution

There are two elements to the path-planning process itself. First, we must know the path we wish to travel, and secondly, we must have a strategy for getting onto and tracking that path. In an indoor environment, the actual paths can be composed of simple straight lines joined by curves. For outdoor road-following robots, on the other hand, paths will tend to be sequences of points that represent undulating curves.

Convergence behaviors

To understand the subtleties of path following, let's consider the simplest possible case, a straight path to a destination. For discussion purposes, we will call the ends of path segments *nodes*.

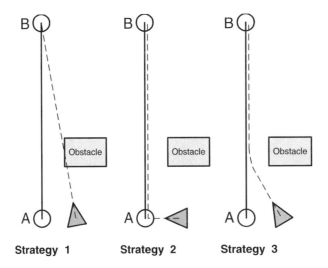

Figure 9.1. Three strategies for getting to a destination

---

[1] See Chapter 5, *Closed Loop Controls, Rabbits and Hounds.*

In Figure 9.1, our robot was supposed to be at Node A, and was to travel from there to Node B. When the robot prepared to execute the move, it realized that its position was actually some distance from where it should have been at Node A. There are three obvious strategies it can use to accomplish the move.

The first possible strategy is to ignore the path, and move straight for destination Node B. The problem with this strategy, at least in this case, is that the robot will run into an obstacle. The fact is that the path was placed where it is to provide a safe avenue for the robot, and when we ignore it we may invite disaster.

The second strategy is to play it safe and first move straight to the correct starting position (Node A) before heading for Node B. This strategy may be safer, but it is also very time consuming. There is also a problem with this strategy if the robot is on the move when it begins the process. In this case, it would need to stop to execute the 90-degree turn to the path.

The third strategy is the "wagon tongue" method, so named because it works like a child pulling a wagon by its tongue. If the child stays on the path, the wagon will smoothly converge onto the path as well. In software, the robot is given a *rabbit* to chase that is on the path, but some *lead* distance closer to the goal than the robot. As the robot chases the rabbit, the rabbit continues to stay the same distance ahead. If the distance to the destination is less than the lead distance, then the strategy will not be appropriate.

So, which strategy is right? The answer is that they all are, at different times and under different circumstances. If the robot is only slightly off the path, and relatively near the destination, then the first strategy is fine. On the other hand, if the robot is grossly off the path and far from the destination, then the second strategy is indicated. For situations between these extremes, the third method is best.

The problem becomes one of definition. How far off the path do we need to be to use the first strategy? In a tight area, the distance will be shorter than in an open area. One solution is to treat paths as objects and provide a property that defines this distance.

Another approach is to create a *path program* that the robot executes to move between the nodes. If we do this, we can include an instruction in this path program that provides a definition of this distance. If our robot is to accomplish specific tasks along the path, then a path program is almost essential.

Closing behaviors

At first, we would think that we could drive to the destination and stop or run to the next destination. Unfortunately, things aren't that simple.

Flashback...

I remember well my first attempt at sending a robot to a node. It was all so simple. My algorithm continually recalculated the vector heading and distance from the robot's position estimate to the destination, and then traveled down that heading until the vector distance was less than half an inch. As it approached the end of the run, it continually recalculated the fastest possible velocity that would still allow it to stop at the destination without exceeding a specified deceleration.

On the first run, the algorithm worked perfectly, so I imprudently[2] called in all my co-workers and attempted to repeat the demonstration. The robot ran purposefully toward its destination, but at the moment when I should have humbly accepted my praise, it all went wrong. The robot abruptly and ungracefully swerved to the left, stopped, turned around, and continued driving in ever-tighter circles like a dog chasing its tail. My algorithm violated the second immutable law of odometry when the robot missed the end point. Once it had stopped, it could not make the minute move required to put it on the target.

As with convergent behaviors, there are several different behaviors that may be required as the robot nears its destination.

Are we there yet?

There are two methods of deciding how much further to drive. The first method is to continually recalculate the distance as we drive. The problem with this comes with the fact that as we approach the end, and the destination is to the side or rear, the vector distance is still positive unless we take into account the vector angle between it and the path.

The second method involves calculating the distance at the beginning of the run, and simply counting down the distance remaining as a parallel process in the odometry. The problem with this approach is that we may end up weaving about for any number of reasons, requiring a longer drive distance than calculated.

---

[2] For a detailed explanation of the phenomenon at work here, see "*The law of conservation of defects and the art of debugging*" in Section 3.

One answer is to use the *parallel* distance to the node to plan the stop, not the vector distance. The second immutable law says we are never exactly where we want to be, and Figure 9.2 shows an extreme case of this. The reasons why our robot could have gotten so far off the straight and narrow are many. It may have had some bad navigation data, and only received good information at the last minute. Or, it could have been forced to circumnavigate an obstacle and has not yet gotten back onto the path.

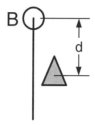

Figure 9.2. Calculating the distance remaining

As the robot approaches the end of the path, it must switch out of the *wagon tongue* behavior and attempt to close on the end point. However, getting exactly to point B is not the only consideration. As in my unfortunate experience, we don't want the robot to swerve near the end of the path as the vector direction to the end node diverges from the path direction due to the robot being to the left or right of the path. Thus, it is usually more important that the robot end the path at the correct heading than at the perfectly correct place.

One solution is to limit the angle between the robot and the path ever more tightly as it approaches the end. Thus, at the end of the path the robot will be consistently oriented parallel to the path. Whatever the lateral position error, we simply accept it. If the robot gets a job to go on forward from Node B, then it can pick one of the *convergence* behaviors already discussed.

There are other considerations in path planning as well. For example, it is common for the robot's collision avoidance to have a specified *stopping* or *standoff distance*. This is the distance from an obstacle that the robot should plan to stop. It is common, however, for the distance from the end of a path to a wall or obstacle to actually be less than the desired *standoff distance*. If this is the case, then the robot may stop before it reaches the end of the path. It may even attempt to circumnavigate the "obstacle."

In another case, the planner may be preparing to execute a tight turn that would be unsafe at the current speed. In fact, the robot needs to have rules about such limits.

Therefore, it is important to recognize that the path planner must communicate with other algorithms such as collision avoidance and speed control to allow the robot to behave intelligently in a wide range of circumstances.

Running on

The situation becomes even more complex if the robot is expected to run on from one node to the next without stopping, as shown in Figure 9.3.

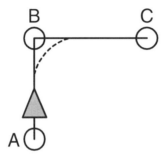

Figure 9.3. Running on past a node

In this case, the robot knows ahead of time that it is not actually running to node B, but is actually arcing past Node B to Node C. In this case, the robot plans the radius of the turn, and this is the distance away from Node B that it must begin to turn. Once into the turn, it steers at a constant rate until the heading is that of the B-C path. When the heading matches, it can revert to a convergence behavior as previously discussed. Remember, the speed in the turn must be slower as the radius is made smaller.

Again, the collision avoidance behavior must interact with the planner to assure the robot takes the appropriate avoidance action depending upon the type of maneuver it is executing.

Bread crumbs and irregular path following

In the case of irregular paths such as roads, the basic techniques just discussed are still valid, but must be modified somewhat. The most common method of describing such paths is as a series (or locus) of points sometimes called "bread crumbs." The term is appropriate as these points are usually saved along a teaching run as the storybook characters Hansel and Gretel used bread crumbs to mark their path back

The best laid plans...

out of the forest. The behaviors already discussed can be adapted to such routes by moving the rabbit from point to point along the path instead of moving it along the straight line of the path.

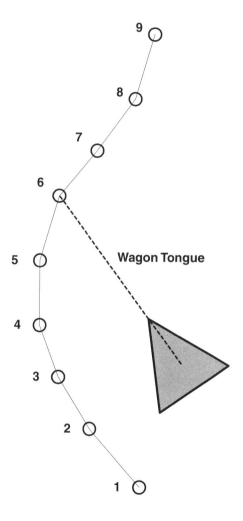

Figure 9.4. Converging onto a bread crumb path

If the distance between points is too small, the number of points can become unnecessarily large, requiring a great deal of memory. Since this path may be transmitted over our communications system, large files are undesirable from a bandwidth standpoint as well. The spacing between points is usually within an order of magnitude of the length of the vehicle, unless they describe a particularly tight geometry.

Smoothest performance can be achieved in following such a path if the rabbit is moved in smaller increments than the distance between the bread crumbs. This can be done by simply performing a straight-line segment between each of the bread crumbs.

The Z axis, maps, and wormholes

For outdoor systems, the z axis must be saved for each of the points in a bread crumb trail if the robot is to operate on anything but a flat surface. Indoors, however, the robot may need to move from floor-to-floor of a building. It is not enough to save the z value of each floor, because the floor plans are typically different at each level. For this reason, instead of a normal z axis, I prefer having the ability to change maps.

In Cybermotion systems, each node has a prefix that is the abbreviation for the map it belongs to. For example, Flr1_AB is node AB on the Flr1 map. Special "paths" known as *wormholes* have nodes that are physically at the same location but that belong to adjoining maps. In this way, an action from Flr1_AB to Flr2_AC creates no executable instructions for the robot, but causes the base station to register that the robot has gone from node AB on the Flr1 map to node AC on the Flr2 map.

An object-oriented approach to the situation is to assign both the value of Z and the map name as properties of the nodes. Paths between nodes would then inherit these properties. Nodes can be objects in their own right that are in turn properties of paths, or they can simply be members of arrays that are properties of paths. The latter representation will conserve memory while the former is more politically correct.

There are compelling reasons to maintain multiple map capability, even in outdoor systems. For one thing, the size of each map can be kept reasonable. Smaller maps mean faster display, pan, and zoom at the base station. In addition, it is quite possible that an autonomous vehicle will need to enter a structure such as a parking garage where multiple maps are essential.

Summary

The behaviors discussed here are but one way of performing path planning and execution. There are undoubtedly many other methods, but the principles and considerations are the same. As the path planner evolves, it will become ever more interconnected with other algorithms and states. Some of these interactions will be discussed in the chapters to come.

The important thing to remember is that each behavior must be capable of operating properly if the robot's position estimate is suddenly changed. If our robot has good solid *live reckoning*, these changes will happen continuously.

CHAPTER 10

Navigation as a Filtering Process

At this point, we have created a way of representing our paths and executing them from odometry alone. This means that we should be able to run simple short programs using only odometry. As it runs without navigational corrections, our robot will become increasingly disoriented, but we can tell that it is doing basically the right maneuvers. This is a distinct advantage in the dual-personality approach as it eases the debug process.

We can think of the path planner and executer as the pilot. It easily follows successive headings for prescribed distances. The navigator, on the other hand, studies the terrain and decides whether the position and heading estimates are correct. If it decides to change them, the pilot's heading commands and distances will change automatically.

Filtering for the truth

A truly capable mobile robot may have many navigation algorithms, each of which is adapted to a certain environment. For example, office buildings usually have hallways with easily detected walls as boundaries. Warehouses, on the other hand, have aisles but these may be defined by nothing more than a paint stripe. Sensors like sonar and lidar can detect the walls, but not the stripes. Cameras can detect both, but only if there is sufficient illumination.

There may also be multiple modes in which a sensor is used. For example, lidar may look for flat surfaces such as walls or special retro-reflectors to provide navigation updates. The algorithms for these two modes are completely different.

Additionally, there may be areas in which it is necessary to specify more than one method of navigation *concurrently*. For example, there may be a warehouse aisle defined by crates along the aisle boundaries. If the crates are removed, walls or reflectors may be exposed behind them. For this reason, our navigator may be looking for multiple features at the same time.

Add to this the fact that almost any feature we are looking for may be mimicked by a false reading, and you begin to see the challenge. For example, a series of boxes along a wall may look just like the wall to our sensor, indicating that we are out of position by the width of the boxes. The navigation process therefore becomes one of selective belief in things that are most likely to be right. As we will see later, most of the time our machine will not believe anything entirely!

There are several metaphors that come to mind when trying to describe this process. In some ways, a good navigator is like a police detective, sorting through clues and contradictions trying to find the truth. The metaphor that I like best, however, is that of an electronic filter that locks in on and passes good data while blocking false information. For optimal performance, this filter will become more and less selective as the platform becomes more and less uncertain of its position estimates.

The importance of uncertainty

Last spring we experienced some terrible flooding. As I dug out from all of the debris I decided to order about 100 tons of rock and gravel to fill in the holes. When the truck driver called with the first load, I gave him instructions on how to find our property. I told him that after the road makes a sharp left turn, he should go about 1.25 miles and find a gazebo on the right-hand side of the road. Just after the gazebo was my driveway. I told him to come across the low-water bridge and drop his load to the right of the driveway just past the bridge.

When he finally arrived, he informed me that there was another gazebo and bridge about half a mile after the turn. Apparently, this gazebo had recently been erected. He had turned into this drive, and was preparing to drop the first 10 tons of rock when a frantic man emerged from the house and wanted to know why he was going to drop the rock on his vegetable garden!

In retrospect, the driver said that it did not seem like a mile and a quarter from the turn to the gazebo, but that the coincidence had been too much for him. The driver had opened his window of acceptance to allow the first gazebo to be taken as the landmark. On the other hand, if he had refused to turn into my driveway because his odometer was only registering 1.2 miles from the curve, then I might never have received my rock. No self-respecting robot would have had an uncertainty window of .75 miles after having driven only .5 miles, but such an uncertainty might have been reasonable, after, say, 20 miles. Thus, uncertainty and the window of acceptance cannot be a simple constant.

For optimal navigation, our filter window needs to be as small as possible and no smaller.

In navigation, we (hopefully) know the true distances and directions very accurately, but we may be uncertain of how far we have traveled, and in exactly which direction. The margin of error is therefore a variable that depends upon how far we have traveled since our last correction. The German bombers, described in Chapter 7, received their only longitudinal correction just before the target and this was for a good reason. By minimizing the distance to be estimated, the margin for error from that point to the target was minimized. The same principle applies for other parameters such as heading.

Uncertainty is thus the gauge of how open we must make our filter to keep from missing our landmarks. The longer the robot runs without a correction, the larger its uncertainty and the higher the risk that it will accept false data.

When we are fortunate enough to acquire valid navigation data, it may only pertain to one relative axis or the heading. After such a fix, we might become more certain of, say our lateral position, but not of our longitudinal position. Although most buildings are relatively orthogonal, this correction may translate to a mix of x and y in the global coordinate system, but that doesn't matter.

Global navigation corrections (such as GPS) apply to all global axes, but do not directly tell us our heading (though we can of course infer this from successive fixes). Many other types of corrections will be with respect to the vehicle's frame of reference. For example, if our robot is going down a hall and using the walls for navigation, our corrections should tell us its lateral position and its heading, but they tell us nothing about its distance traveled. In this case, our uncertainty in the corrected axes will diminish, but the uncertainty in the uncorrected axes will continue to accumulate. The solution is to treat all corrections as being relative to our frame of reference. Thus, a GPS fix is treated as both a lateral and a longitudinal correction.

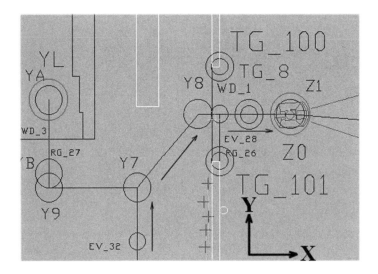

Figure 10.1. Correcting multiple axes
(Courtesy of Cybermotion, Inc.)

In Figure 10.1, the robot has driven from the bottom of the screen through the nodes Y7, Y8, and then stopped at node Z1. As it approached Y7, it collected sonar data (shown as "+" marks) from the wall on its right. When this data was processed, the "fix line" was drawn through the "+" marks to show where the robot found the wall.

From this fix the robot corrected its lateral position and its heading. As it turned the corner and went through a wide doorway, it measured the edges of the door and drew a "+" where it found the center should have been. Again, it made a correction to its lateral position, but while the first lateral correction translated to a correction along the x-axis of the map, this correction corrected its error along the y axis. The uncertainty of the corrected axes will have been reduced because of each of these fixes.

Modeling uncertainty

To calculate uncertainty is to model the way the robot's odometry accumulates error. For the purposes of this discussion, I am going to assume a platform that can turn in place. If you are working with a drive system that cannot do this, such as an Ackerman system, then you will need to interpret some of these concepts to allow for the coupling between driving and steering.

Uncertainty is most easily calculated relative to the vehicle's frame of reference. The reasons for keeping uncertainty in a relative reference are several. Most importantly, errors will occur for different reasons along the vehicle's lateral and longitudinal axes. Heading errors will occur for different reasons yet, and it will induce lateral error as the robot drives.

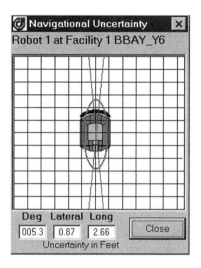

Figure 10.2. Uncertainty diagnostic display
(Courtesy of Cybermotion, Inc.)

Figure 10.2 shows a diagnostic display for monitoring uncertainty. The ellipse indicates the lateral and longitudinal uncertainty, while the intersecting lines indicate the heading uncertainty. In this case, the uncertainty represents the absolute boundary for correction data that is received. Data outside of this boundary will be ignored, but this does not mean that a correction of slightly less than this magnitude would be used to correct the position estimates directly. More will be said about this in the coming chapters.

Azimuth (heading) uncertainty

A robot that is not moving generally accumulates no error[1]. Heading uncertainty is therefore caused primarily by the acts of turning or driving. Like other forms of uncertainty, this one has both an inherent platform component and a surface component.

---

[1] Actually, it can falsely accumulate uncertainty if an encoder repeatedly registers forward and backward ticks as the result of Servo-hunting or edge jitter on an encoder. Therefore, it is a good policy to incorporate hysteresis on all such signal inputs.

If a robot is attempting to turn in place on a loose throw rug, the error may be worse than on a tile floor. Generally, the platform component of heading uncertainty for steering can be calculated by simply multiplying the cumulative magnitude of a turning action by an error factor (in units of degrees of error per degree of turning). It is desirable that this factor be readily changed by the robot. It may be changed in the initialization sequence, a path program, or a learning program of the platform as necessary. These error factors are a prime example of a good use of blackboard variables.

If the mobile platform has a steering encoder that causes tick interrupts, the calculation becomes nothing more than adding the error factor to the accumulated uncertainty in a register.

If the mobile platform does not have an interrupt for steering, then it is necessary to do this calculation during the drive interrupts. This can be done by noting the amount that the steering angle has changed since the previous interrupt. This calculation must be done before the lateral uncertainty can be calculated.

Similarly, the heading uncertainty generated by driving is a factor of the distance moved. Every time the odometry advances the robot's position estimate by one tick, it can add an appropriate amount to the heading uncertainty accumulator.

Generally, the error factors are quite small numbers, so it is convenient for the uncertainty accumulator to be two or more words. The incremental increases are then added to the lower word with carry to the higher register(s). The higher register(s) are the only ones accessed by consuming tasks. In this way, the math for these calculations can stay in integers and therefore execute very fast.

Heading errors related to driving, and thus uncertainty, are generally symmetric. Turning errors are, however, not generally symmetric. A turn is much more likely to come up short than long. The ideal model would take this into account, but very decent performance can be accomplished by treating the uncertainty as if it were symmetric by using its worst case factor.

Note that while we are performing uncertainty calculations using simple multiplication and addition, the effect of doing this iteratively is to produce the *integral* of these functions.

Longitudinal uncertainty

Longitudinal error accumulates primarily from slippage, wheel wear, and improper tire inflation. There are again two elements to be taken into account: errors from the mobile platform itself, and those from the surface it is running on. A robot running in snow will likely under-run the distance it thinks it has traveled.

Longitudinal uncertainty is thus nothing more than the distance traveled, multiplied by an error factor. Longitudinal uncertainty is not generally symmetric. On level ground, a robot may tend to underrun the path but will seldom overrun it. On an uphill slope, this tendency is more significant yet. On a downhill slope, the robot will tend to overrun its path if slippage occurs during braking. Again, reasonable results can be gained with a symmetric assumption about longitudinal uncertainty, but the advantage of keeping separate values for the two directions is fairly strong.

As the robot turns, the accumulated lateral and longitudinal uncertainty are simply converted into a vector, rotated, and converted back into lateral and longitudinal components. If a robot turns 90 degrees, then the lateral uncertainty will become the longitudinal uncertainty and vice versa.

Lateral uncertainty

Lateral error is predominantly accumulated as the result of driving with a heading error. Thus, lateral uncertainty is accumulated by multiplying the distance traveled by the sine of the heading error. If this accumulation is done frequently, the fact that the heading error is not constant can be ignored. Since heading can be approximated as being symmetric, so can the accumulated increments of lateral uncertainty.

Reducing uncertainty

Up until now, we have been mostly concerned about accumulating uncertainty. Clearly there must also be a way of diminishing it. When an axis accepts a fix, its uncertainty should also be reduced. But beware, this process is not as simple as zeroing the accumulator for the axes that were corrected.

First, if we allow uncertainty to go to zero, we will not be able to accept any new fixes until we have traveled enough to accumulate new uncertainty. Since there is always a bit of error in sensor readings, building measurements, and other factors, we need to establish a *minimum uncertainty threshold* below, which we do not go.

Let's assume that we have just received a correction of the lateral position of the robot. No matter how good our filtering is, there is always a finite chance that this correction is not valid. If we reduce our uncertainty too much, we may lock our filter into this correction and refuse to believe the correct data when we next receive it.

Therefore, uncertainty should never be reduced below the magnitude of the current correction or the minimum uncertainty threshold, whichever is larger.

Learning to be accurately uncertain

If our uncertainty estimates are too pessimistic, they will potentially allow bad data to be accepted into the position estimates. If the uncertainty estimates are too optimistic, then they may cause true navigational corrections not to be believed. Thus, the better the uncertainty model, the tighter our filter, and the better the navigation should be.

Unfortunately, things change. We may find the perfect parameters for uncertainty at installation, only to find them less accurate months or years later. Gears wear, carpet wears, road conditions change, etc. For this reason, it is very useful to learn the proper error factors as the robot drives. This learning can most easily be accomplished by a program that statistically studies the operation of the robot and the fixes it accepts. If these statistics indicate that corrections are beginning to regularly approach the magnitude of the uncertainty, it may be useful to learn new error factors.

Uses of uncertainty

The importance of the uncertainty estimate cannot be overstated. In the coming chapters, we will see that this estimate is crucial in acquiring and processing navigation data. It is also important in controlling the behavior of the robot in such a way as to assure safe operation.

11

Hard Navigation vs. Fuzzy Navigation

We have discussed the importance of knowing uncertainty as a control over the navigation filtering process, but we have yet to consider the mechanism for this control. Here we leave the world of rigid mathematics and move into a more subjective one. The concepts that follow are ones that I have developed and found useful. Following the principle of *enlightened instinct* you may find methods that work even better. These proven concepts are very simple and straightforward[1].

Sensor data and maps

Sensors represent one of the many *enabling technologies* that are making autonomous robots more capable every year. A decade ago, the best affordable sensor for a mobile robot was a short-range sonar system[2]. Navigation based on such sensors worked very much the way a blind person uses a cane.

Today, we have affordable lidar sensors that can read ordinary objects beyond 10 meters, and reflective markers as far away as 50 meters. Despite these developments, the fundamental problem remains the same. At any given time, a robot can image only a small part of its world with its sensors, and the images received will always contain anomalies.

---

[1] Nothing more complex than simple vector manipulation is required. (See Appendix A.)

[2] Even with such limited sensors, robots were developed that could navigate many indoor environments.

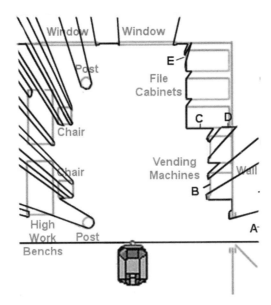

Figure 11.1. Lidar range data for an indoor environment

Range data from a lidar is far richer than that from sonar, but even so, it remains a skewed and partial perception of reality as Figure 11.1 demonstrates. For example, areas like "A" where no target is present are usually reported as being at the system's maximum range. Shiny objects such as the glass fronts of the vending machines "B" *specularly* reflect the beam instead of returning it to the ranging unit. *Specular* reflection is the effect seen in a mirror, with the angle of incidence being equal to the angle of reflection.

In the case of the upper vending machine, the lidar actually reports the reflection "D" of the file cabinet "C" because of these reflections. The shiny paint of the upper file cabinet also causes a small amount of specular reflection "E" because the angle of incidence of the laser beam is so shallow. Notice that the workbenches on the left are above the lidar scan height, so only their legs and those of their chairs are detected by the lidar. Worse, some of these legs are masked by other objects.

Figure 11.2. Lidar range data after initial filtering

Range data from Figure 11.1 can be filtered by simple techniques such as eliminating any lines that lay radial to the sensor or which have one or more end points at the range limit. The result is shown in Figure 11.2. This scene still bares little resemblance to the map. As the robot moves, it will see things differently, and if the odometry is sufficiently reliable, a more complete map can be built from successive perspectives.

Here we face the same tradeoff discussed earlier. If we wait and drive long enough to see the room from many perspectives, our position estimate may have degenerated to the point that we are no longer correlating things in their proper places. Therefore, the robot will quickly need to identify things in the scene it can use to verify and correct its position estimate. One solution is to select navigation *features* ahead of time that will be easily recognized.

Navigation features

A navigation *feature* is an expected (permanent) object that a robot's sensors can detect and measure, and that can be reasonably discriminated from other objects around it. This navigation feature may be a part of the environment, or something added to the environment for the robot's purposes[3]. Features that are added to the environment for navigation purposes are called *fiducials* or *fids*. Robot engineers often refer to the need to add such artificial references as "decorating the environment," a phrase of mild derision.

In the days when sonar was the best available sensor technology, the programmer had no choice but to use the very narrow range of navigation features that sonar could detect and discriminate. As sensor systems have become more capable, the navigation features available to robots have become more complex. In many ways, this development cycle has paralleled the history of modern aircraft navigation, especially as it accelerated rapidly during World War II.

Here we pick up the story where we left off in Chapter 7. As WWII progressed, the allies turned the tables and began bombing operations over occupied Europe. Now the British Royal Air Force (RAF) began experiencing the same heavy losses of bombers that the Luftwaffe had experienced during the Battle of Britain. Like the Germans, they switched to night bombing, and they too found accurate navigation over blacked-out terrain to be very difficult. Since they had defeated the German beam systems, the British looked for a more robust solution to the navigation problem.

A new downward-looking radar was developed that could show the navigator a representation of the ground below. Although much of the terrain showed up as a confusing jumble of blips (like the lidar data in Figure 11.2), some features like rivers and shorelines stood out clearly. The British thus navigated by dead reckoning until such an easily recognized feature was reached, and then corrected their position estimates. Missions were planned to assure that the bombers would fly over an easily recognized feature shortly before they reached their targets[4].

---

[3] Even data from a GPS receiver is essentially the result of imaging a set of satellites. Most of the concepts discussed here will thus work if a GPS system is simply considered to be one of the robot's sensors.

[4] In response to this system, the Germans developed a receiver for their night fighters that could home in on the radar transmissions of the RAF bombers. This continual game of measures and countermeasures spurred the ever-faster development of electronic technology. Developments ranging from microwave ovens to computers and fax machines all have their roots in this incredible period of history.

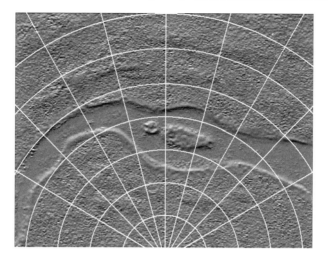

Figure 11.3. Radar image of islands in a river as a navigation feature

Hard navigation

Navigation of ships and planes was originally accomplished using manual calculations. These calculations took some considerable time, so updates in the craft's position estimates were infrequent. When a reasonable correction was calculated, it was simply accepted as fact. This is an early example of what we will call *hard navigation*.

The first semi-autonomous wheeled vehicles were AGVs (automatic guided vehicles) that followed physical paths (stripes or wires in the floor). These vehicles performed no true navigation as they had no sense of their position other than as related to the path they were following. As these systems began to evolve toward more flexibility, the first attempts at true navigation quite naturally used hard navigation.

One very early laser navigation system developed by Caterpillar Corp. triangulated between reflective targets to fix its position. These targets looked like barcodes on steroids, being several feet high and at least a foot wide. The reasoning for this was that the vehicle needed to know which reflector was which so that it could perform its calculations without the possibility of ambiguity. The vehicle could find its position by no other means than this calculation, and it "believed" each calculation literally. This is an example of hard navigation. Even the first robots that we field-tested at Cybermotion used hard navigation.

Flashback...

One of the problems of autonomous robot development is that our creations will quickly evolve to work well in the development environment, but may be confounded by other environments[5]. The first robot we fielded in a real-world application (circa 1987) had only a few navigation tricks. It could home in on a beam from its charger to become oriented (*referenced*) and it could use wide-beam digital sonar to image wall surfaces and correct its odometry. In our facility this, of course, worked wonderfully.

When, however, we turned the robot loose in the customer's facility, it weaved down the halls like a drunken sailor. The customer was not impressed. This happened on a Friday, and our customer was scheduled to show the system to his vice president on Monday. It would be a busy weekend.

We determined the cause of the problem rather quickly. Where the walls in our offices had been flat, the walls in this new facility were interrupted frequently by recessed picture windows. There were also large doors that were routinely left slightly ajar. The problem occurred when the robot imaged these sections. The heading correction implied by these features was often at a significant angle to the true heading of the hallway. Each time such a correction was made the robot would change course, only to correct itself a few meters further down the hall.

The first thing we did was to set the acceptance threshold on the heading correction much tighter[6]. If a fit wasn't close to our current heading, we would simply reject it. This eliminated the false angular corrections, but then other places around the building began to cause trouble.

There were several places where the robot was forced to drive for some distance without any navigational corrections. When the robot crossed these gaps and began getting wall readings again, it often rejected them as being at too great an angle.

Over the weekend, we applied the first crude fuzzy logic to our navigation process. The results were remarkable, and we got through the demonstration that following Monday with flying colors.

---

[5] The mechanism at work here is clearly parallel to natural selection. This is another example of how closely robotics parallels nature.

[6] We could change this threshold on site because it was an easily accessible blackboard variable as discussed in Chapter 6.

The concept of fuzzy navigation

There is an old saying, "Don't believe anything you hear, and only half what you see!" This could be the slogan for explaining fuzzy navigation. When police interrogate suspects, they continue to ask the same questions repeatedly in different ways. This iterative process is designed to filter out the lies and uncover the truth.

We could simply program our robot to collect a large number of fixes, and then sort through them for the ones that agreed with each other. Unfortunately, as it was doing this, our robot would be drifting dangerously off course. We need a solution that responds minimally to bad information, and quickly accepts true information.

The trick is therefore to believe fixes more or less aggressively according to their *quality*. If a fix is at the edge of the believable, then we will only partially believe it. If this is done correctly, the system will converge on the truth, and will barely respond at all to bad data. But how do we quantify the quality of a fix? There are two elements to quality:

1. Feature image quality
2. Correction quality

Feature image quality

The image quality factor will depend largely on the nature of the sensor system and the feature it is imaging. For example, if the feature were a straight section of wall, then the feature *image quality* would obviously be derived from how well the sensor readings match a straight line. If the feature is a doorway, then the image data quality will be based on whether the gap matches the expected dimensions, and so forth.

The first level of sensor processing is simply to collect data that could possibly be associated with each feature. This means that only readings from the expected position of the feature should be collected for further *image* processing. This is the first place that our *uncertainty* estimate comes into use.

Figure 11.4 shows a robot imaging a column. Since the robot's own position is uncertain, it is possible the feature will be observed within an area that is the mirror image of the robot's own uncertainty. For example, if the robot is actually a meter closer to the feature than its position estimate indicates, then to the robot the feature will appear to be a meter closer than expected. The center of the feature may thus be in an area the size of the robot's uncertainty around the known (programmed) position of the feature. Since the column is not a point, we must increase this area by the radius of the column and collect all points that lay within it for further processing.

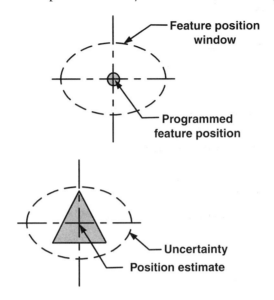

Figure 11.4. Using position uncertainty to find a *feature's* window

The feature position window gets even larger when we include the robot's heading uncertainty as demonstrated in Figure 11.5.

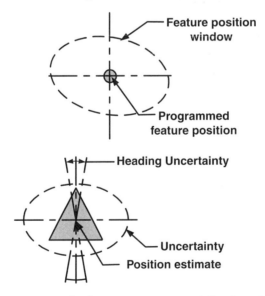

Figure 11.5. Feature window compensated for heading uncertainty

Since the calculations to determine if any given point lies inside a rotated ellipse are relatively time consuming, it is usually adequate to approximate the window into a heading and range as shown in Figure 11.6. In this example, the robot's uncertainty is relatively large, and it is obvious that this is going to open the window for a lot of data that may not be associated with the feature we are looking for. A larger window means more processing of unrelated data, and a greater likelihood of a false identification of a feature.

The lower the uncertainty, the better the navigation process works, and thus the lower the uncertainty should subsequently become. The philosophy we will be developing is one of layered filtering, with each layer being relatively open. It is also an iterative process that closes in on the truth while largely rejecting invalid information.

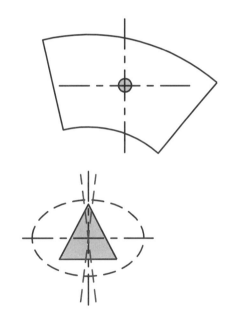

Figure 11.6. Range and heading window approximation

Sensor processing should next screen the data to attempt to enhance the quality. For example, if a laser sensor takes 100 readings along a wall, it may well have data from objects on the wall such as conduit, thermostats, moldings, and so forth. If it uses a linear regression to find the line represented by the readings, and then goes about throwing out points until the RMS error of the remaining points is below a threshold, then it will probably have pretty good data to send to the next level of processing.

The number of points remaining that represent a flat surface is one indication of the image quality in this example. For example, if we end up with 98 out of 100 points, then we have a very high confidence in the data. However, if we end up with, for example, 10 points, there is a chance that these points are not really the surface, but just ten random points that lie on a straight line. The average error of the remaining points along the wall is another indicator of quality. If we throw out 25 points, but the remaining 75 points are almost perfectly along a line, then this is a positive quality indicator.

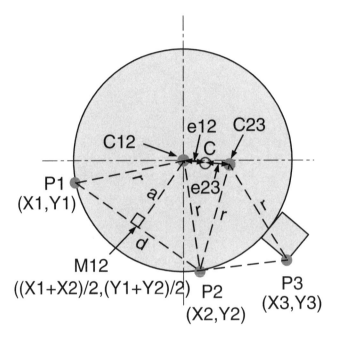

Figure 11.7. Calculating image quality for a pillar

Instead of a flat surface, let's approach a more interesting feature. Figure 11.7 shows a large building pillar that has been imaged by a lidar system. In this example, the pillar is a considerable distance from the robot, so pre-filtering has found only three range points—P1, P2, and P3—that are in the expected vicinity of this feature and may therefore belong to it. We know the radius "r" of the column from the robot's program, and we know where the column should be, so we begin by processing adjoining pairs of points.

First, we calculate the vector between P1 and P2, and find its magnitude, which we will call "*d*". We assume that the points P1 and P2 lie on the surface of the column and then we can calculate where these points imply that the center of the column is. We find this by adding the vector "*a*" to the midpoint of the line from P1 to P2. It is relatively straightforward to find the magnitude of "*a*" since we know all three sides of an equilateral triangle. By treating half of this triangle as a right triangle, and using the formula that the square of the hypotenuse is equal to the sum of the squares of the sides, we can quickly show that:

$$a = \text{sqrt} \ (\ r^2 - (d/2)^2 \ ) \qquad \qquad \text{(Equation 11.1)}$$

We also know that the azimuth of vector "*a*" is 90 degrees counterclockwise from the vector azimuth for P1-P2. By adding the vector "*c*" to the midpoint of P1-P2, we find the center of the column "C12" as *implied* by points P1 and P2. Then we repeat the process for points P2 and P3. Since P3 does not lie on the column, the implied center C23 is not at the same place as C12, but our robot does not yet know which point is wrong. We average the two implied centers and get the center position "C." The error for the P1-P2 implied center is thus the distance e12, and that for P2-P3 is e23.

Had there been additional points for the feature, we would have had enough data to tell that a point like P3 was not a good reading because the implied center of segments using P3 was further from the average center than the implied centers of other point pairs. We could have eliminated P3 from the collection. After removing the bad points, we can recalculate the implied centers for the neighbors on either side of them, as well as the average center.

Once we have filtered down to the points we wish to keep, the detected center of the column is taken as the average of the remaining implied centers[7]. However, what is the quality of this feature data? This is where the *enlightened instinct* kicks in. One frame of reference is the radius of the columns. If the implied centers vary by a significant fraction of the radius of the column, then the image really doesn't look much like the column we are looking for.

---

[7] I prefer simple averages to more complex statistical calculations because they are very fast to calculate and the difference in results is usually insignificant if the points have been filtered for outliers.

For reasons to be explained shortly, the quality factor should be a number between zero and one. If we call the average magnitude of the errors of the implied centers E, then we could use the simple formula:

$$Q_i = (r - E) / r \qquad \text{(Equation 11.2)}$$

Here "r" is the known radius of the column. For E equals 0, the quality is always 1. If the column radius was 4 meters and the average error of the implied centers was 1 meter, we would get an image quality factor of 75%. For E greater than r, the Q factor would be considered zero and not negative. The better our quality factor represents the true validity of a feature, the more quickly the robot will home in on its true position, without being fooled by false sensor readings.

For relatively small poles, the above formula may produce artificially lower quality numbers because the resolution of the ranging system and its random noise errors become significant compared to the radius of the column. In such cases, we might want to replace the radius with a constant that is several times larger than the typical ranging errors. For reflective fiducials, the discrimination of the lidar sensor is adequately good that an image quality of unity may be assumed.

The goal is simply to achieve a *relative* representation of the quality of the image data. To this end, we must also consider whether there were enough points to be reasonably confident of the validity of the implied centers. We might arbitrarily decide that five good points or more are okay. For fewer points, we could reduce the image quality calculated above—by, say 25% for each point less than 5. For two points, we have no cross check at all, so we will assign the column a 25% quality.

It is important to realize that the robot's program would simply specify that it was to use a column feature of a certain radius, and at a certain expected location. This command in the robot's program would then invoke the search for and processing of the column data. Unfortunately, our robot cannot determine its position from a single-circular feature like a column. For this it would need to see at least two such features.

The first thing we can do with the image quality is to compare it to a threshold and decide whether the image is good enough to use at all. If it is above the minimum threshold, we will save the calculated center and the quality for later use in our navigation process.

Correction quality (believability)

Any good salesman can tell you that it is easier to convince a person of something that is close to their belief system than of something that is outside of it. Likewise, our robot will want to believe things that confirm its existing position estimate. The problem is that we cannot apply a simple fixed threshold around the current estimate, as explained in our example of the drunken robot. This is the second place where our *uncertainty* estimate comes in.

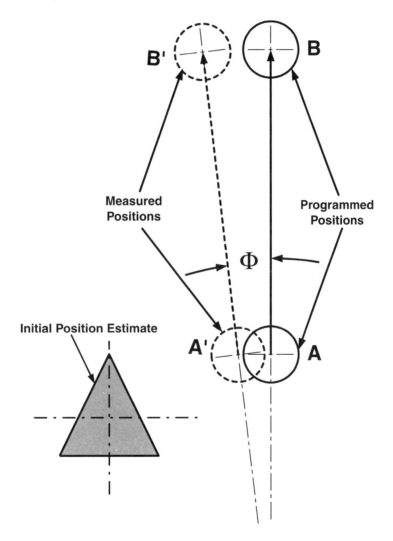

Figure 11.8. Calculating heading error from two columns

The robot's calculated *uncertainty* is the metric against which implied corrections are compared. The process is then one of simple trapezoidal fuzzy logic as described in Chapter 4.

In Figure 11.8, the robot has imaged a pair of columns that it expected to find at positions A and B. The implied centers calculated from the range measurements were found to be at A′ and B′. It is a simple matter to calculate the hard heading correction −Φ that this data implies. However, if we take this whole correction, we may actually add more error to our heading rather than reducing error. If the implied centers of

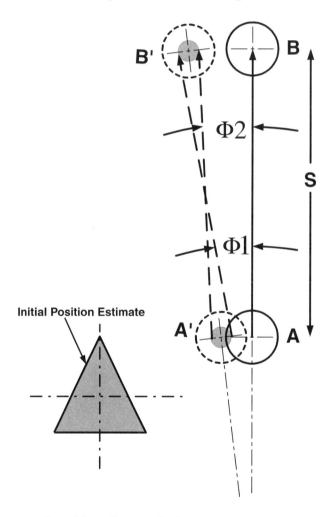

Figure 11.9. Calculated heading variation due to implied center variation

the columns varied only 40% of the radius, as shown by the shaded circles in Figure 11.9, then the variation this would imply for the heading correction could range from $-\Phi1$ to $-\Phi2$. Notice that the $-\Phi1$ correction is nearly twice the calculated $-\Phi$ in Figure 11.8, while the $-\Phi2$ correction is nearly zero.

Both $\Phi1$ and $\Phi2$ are in the same direction, so there is likely to be some truth in their implications, we just don't know how seriously to take them. We can apply other checks as gatekeepers before we must decide how much of the correction $-\Phi$ we want to apply to our odometry. We have already calculated the image quality of each column, and rejected the use of any columns that were not above a reasonable threshold. We can also check the distance between the implied centers of the columns and make sure it is close to the programmed spacing (S).

Now the time has come to decide the quality of the heading correction implied by these ranges. To do this we calculate the heading correction believability or *azimuth quality* of the implied correction. The azimuth correction quality is:

$$Q_{AZ} = (U_{AZ} - |\Phi|) / U_{AZ} \qquad \text{(Equation 11.3)}$$

Here U_{AZ} is the current azimuth (heading) uncertainty. Note that as the implied correction becomes small with respect to the heading uncertainty, the quality approaches unity. If the implied correction is equal to or greater than the azimuth uncertainty, the quality is taken as zero.

Again, there are two reasons we might not believe this whole heading correction: first, because the data that created it was not clean (or there was not enough of it to know if it was clean); and secondly, because the correction did not seem reasonable given the robot's confidence in its current heading estimate. Therefore, we can reduce the amount of an implied correction by simply multiplying it by the various quality factors that affect it. If Q_A is the image quality for column A and Q_B is the image quality for column B, then *fit quality* for the whole implied azimuth correction is:

$$Q_{FIT} = Q_A * Q_B * Q_{AZ} \qquad \text{(Equation 11.4)}$$

The correction Φ_{COR} to be used on this cycle will be:

$$\Phi_{COR} = - (Q_{FIT} * \Phi) \qquad \text{(Equation 11.5)}$$

To understand how simply this works, consider a robot imaging these columns and returning a low azimuth quality because the implied correction is nearly as large as its azimuth uncertainty. For argument's sake, let's say that the image quality is very good (nearly unity). On the first set of ranges, the vehicle will take only a small correction. If, however, that correction was valid, and the next set of ranges is in general agreement with the first, then the azimuth quality of this second calculation will be slightly higher and the robot will take a more significant proportion of the implied correction. If this trend continues for a few cycles, the robot will begin to aggressively believe most of the corrections implied by the sensor. If the first correction had been invalid, however, the robot would not have significantly degenerated its position estimate.

If our robot can observe other columns or landmarks at the same time, then deciding which set of columns to use can be as simple as comparing the fit qualities and taking the best pair. A more exacting approach is to use fuzzy democracy and allow all the implied corrections a vote proportional to their calculated qualities. For example, if there are two reasonable implied corrections Φ_{AB} and Φ_{BC} with fit qualities $Q_{FIT\text{-}AB}$ and $Q_{FIT\text{-}BC}$, then the correction could be calculated as:

$$\Phi_{COR} = ((Q_{FIT\text{-}AB} * \Phi_{AB}) + (Q_{FIT\text{-}BC} * \Phi_{BC})) / (Q_{FIT\text{-}AB} + Q_{FIT\text{-}BC)} \quad \text{(Equation 11.6)}$$

Also note that the larger the spacing S between columns, the better the resolution on heading will be for any given measurement noise in the implied centers of the columns. A better representation of the image quality of a pair of columns might consider this.

The nice thing about fuzzy navigation, however, is that it is not necessary to precisely quantify the quality factors; if they are reasonable, the system will converge on the proper solution very quickly. If our system is converging too slowly, then most likely our quality factors are too stringent. If it is over-correcting, then our quality factors are probably too generous.

We have yet to correct the lateral and longitudinal axes. These implied corrections will be equal to the apparent displacement of the features after compensating for the azimuth displacement. We will use the closest feature (A), because the effect of any error in the measured azimuth displacement will be smaller. To compensate for the heading error, we find the vector from the center of our position estimate to the measured center of the column A'. We simply rotate this vector by the implied azimuth error, but in the opposite direction, and then add it back to our position estimate and the result is the compensated position A''. The displacement of A'' from the programmed position A is our full implied position error, consisting of a lateral and a longitudinal component. For simplicity, we will refer to lateral dimen-

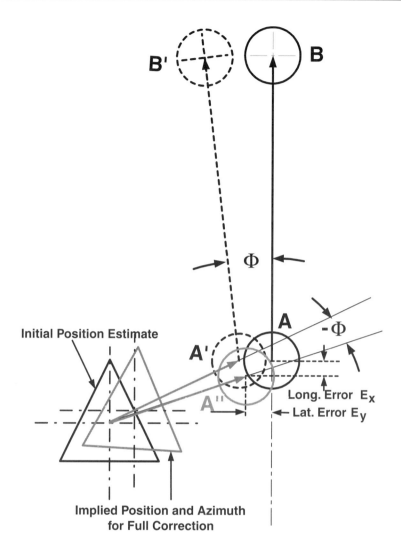

Figure 11.10. Correcting the X and Y axes

sions as lowercase "x," and longitudinal dimensions as lowercase "y." The lowercase designates that the coordinates are relative to the robot's center and direction.

The question now is how much of this implied correction should we use? We will determine this in much the same manner that we determined the heading correction. The apparent position of column A will have two error factors: the robot's position error and the robot's heading error.

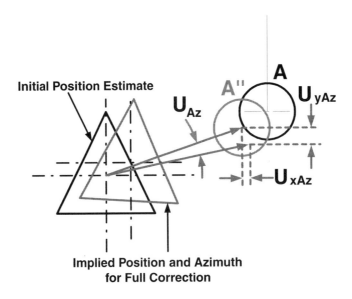

Figure 11.11. Apparent position uncertainty as a result of azimuth uncertainty

Figure 11.11 demonstrates how we might determine the error in the apparent position of the column that *could* be contributed by azimuth uncertainty. We simply take the vector from the robot's current position estimate to the compensated measured position A'' and rotate it by the azimuth uncertainty (U_{Az}). The x and y position differences between the two vectors' ends (U_{xAz} and U_{yAz}) are the possible x and y errors that could be attributable to the azimuth being incorrect.

Unfortunately, this observational uncertainty may add to or subtract from the apparent position error caused by the robot's lateral and longitudinal uncertainty (its true position error). However, we can estimate the likelihood that an observed error was the result of the vehicle's position error vs. its heading error. We will call these factors the *observational* quality for the lateral and longitudinal corrections.

$$Q_{Ox} = (U_x - U_{xAz}) \,/\, U_x \qquad\qquad \text{(Equation 11.7)}$$

And

$$Q_{Oy} = (U_y - U_{yAz}) \,/\, U_y \qquad\qquad \text{(Equation 11.8)}$$

Where:

U_x = Platform lateral uncertainty

U_y = Platform longitudinal uncertainty

Again, Q factors less than zero will be taken as being zero quality. As the observational uncertainty approaches the platform uncertainty for either axis, the quality of the observation for that axis approaches zero. Notice that for the example in Figure 11.11, the lateral uncertainty of the observation that resulted from azimuth uncertainty is quite small compared to the corresponding longitudinal component. This means that we can have more faith in the lateral observation.

Now we consider the most fundamental Q factor, and that is just how believable the implied correction for an axis is compared to the uncertainty for the same axis.

$$Q_x = (U_x - |E_{xAz}|) \,/\, U_x \qquad\qquad \text{(Equation 11.9)}$$

And

$$Q_y = (U_y - |E_{yAz}|) \,/\, U_y \qquad\qquad \text{(Equation 11.10)}$$

Where:

E_x = Observed lateral error (see Figure 11.10)

E_y = Observed longitudinal error

If the absolute value of an implied correction for an axis is greater than the uncertainty for that axis, then we will simply discard the correction. We now have two quality factors for each axis: the observational quality and the believability. The fit quality for each axis as follows:

$$Q_{FITx} = Q_x * Q_{Ox} \qquad\qquad \text{(Equation 11.11)}$$

And

$$Q_{FITy} = Q_y * Q_{Oy} \qquad\qquad \text{(Equation 11.12)}$$

The correction we will actually make for each axis is calculated as it was for azimuth, by multiplying the full implied correction (error) by the Q factor for the whole fit.

$$x_{COR} = -(Q_{FITx} * E_x)$$ (Equation 11.13)

And

$$y_{COR} = -(Q_{FITy} * E_y)$$ (Equation 11.14)

Assume that this robot had begun this whole process with significant uncertainty for both azimuth and position, and had suddenly started to get good images of these columns. It would not have corrected all of its axes at the same rate.

At first, the most improvement would be in the azimuth (the most important degree of freedom), then in the lateral position, and then, somewhat more slowly, the longitudinal position. This would be because the corrections in the azimuth would reduce the azimuth uncertainty, which would increase the observation quality for the position measurements. However, we have only discussed how uncertainty grows, not how it is reduced.

Other profiles

The equations discussed thus far are the most primitive possible form of fuzzy logic. In fact, this is not so much a trapezoidal profile as a triangular one. Even so, with the correct constants they will produce very respectable results.

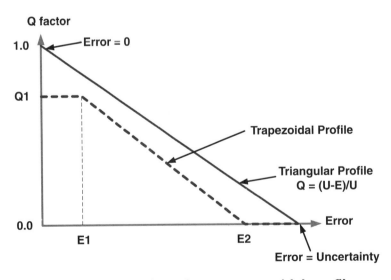

Figure 11.12. Triangular vs. trapezoidal profile

Figure 11.12 shows another possibility for this function. In this case, we simply refuse to grant a quality factor greater than Q1, so the navigation will never take the whole implied correction. Furthermore, any error that is greater than E2 is considered so suspect that it is not believed at all. If the breakpoints of this profile are stored as blackboard variables, then the system can be trimmed during operation to product the best results.

The referenced state

It should be obvious that the navigation scheme we are pursuing is one that requires the robot to know roughly where it is before it can begin looking for and processing navigation features. Very few sensor systems, other than GPS, are global in nature, and very few features are so unique as to absolutely identify the robot's position.

When a robot is first turned on, it is *unreferenced* by definition. That is, it does not know its position. To be referenced requires the following:

1. The robot's position estimate must be close enough to reality that the sensors can image and process the available navigation features.

2. The robot's uncertainty estimate must be greater than its true position error.

If our robot is turned on and receives globally unique position data from the GPS or another sensor system, then it may need to move slightly to determine its heading from a second position fix. Once it has accomplished this, it can automatically become *referenced*.

If, however, a robot does not have global data available, then it must be told where it is. The proper way to do this is to tell the robot what its coordinates and azimuth are, and to also give it landmarks to check. The following incident will demonstrate the importance of this requirement.

Flashback...

One morning I received an "incident report" in my email with a disturbing photo of a prone robot attached. A guard had stretched out on the floor beside the robot in mock sleep. Apparently, the robot had attempted to drive under a high shelving unit and had struck its head (sensor pod) on the shelf and had fallen over. This was a very serious—and thankfully extremely rare—incident.

Those who follow the popular TV cartoon series "The Simpsons" may remember the episode in which Homer Simpson passed down to his son Bart the three lines that had served him well his whole life. The third of these lines was, "It was that way when I found it, boss." In the security business, an incident report is an account of a serious event that contains adequate detail to totally exonerate the writer. In short, it always says, "It was that way when I found it, boss."

This incident report stated that the robot had been sent to its charger and had mysteriously driven under the shelving unit with the aforementioned results. In those days we had a command line interface that the guard at the console used to control the robot. A quick download of the log file showed what had happened.

Instead of entering the command, "Send Charger," the operator had entered the command "Reference Charger." This command told the robot its position was in front of the charger, not in a dangerous aisle with overhead shelves. The reference program specified two very close reflectors for navigation. The robot was expected to fine tune its position from these reflectors, turn off its collision avoidance, and move forward to impale itself on the charging prong. Since there was no provision in the program for what to do if it did not see the reflectors, the robot simply continued with the docking maneuver.

The oversight was quickly corrected, and later interlocks were added to prevent the repetition of such an event, but the damage had been done. Although this accident resulted in no damage to the robot or warehouse, it had a long-lasting impact on the confidence of the customer's management in the program.

If the robot's uncertainty estimate is smaller than the true error in position, then the navigation agents will reject any position corrections. If this happens, the robot will continue to become more severely out of position as it moves. At some uncertainty threshold, the robot must be set to an *unreferenced* state automatically.

If the robot successfully finds the referencing features, then—and only then—can it be considered *referenced* and ready for operation. Certain safety related problems such as servo stalls should cause a robot to become *unreferenced*. This is a way of halting automatic operations and asking the operator to assure it is safe to continue operation. We found that operators ranged from extremely competent and conscientious to untrainable.

Reducing uncertainty

As our robot cavorts about its environment, its odometry is continually increasing its uncertainty. Obviously, it should become less uncertain about an axis when it has received a correction for that axis. But how much less uncertain should it become?

With a little thought, we realize that we can't reduce an axis uncertainty to zero. As the uncertainty becomes very small, only the tiniest implied corrections will yield a nonzero quality factor (Equation 11.9 and 11.10). In reality, the robot's uncertainty is never zero, and the uncertainty estimate should reflect this fact. If a zero uncertainty were to be entered into the quality equation, then the denominator of the equation would be zero and a divide-by-zero error would result.

For these reasons we should establish a blackboard variable that specifies the minimum uncertainty level for each axis (separately). By placing these variables in a blackboard, we can tinker with them as we tune the system. At least as importantly, we can change the factors on the fly. There are several reasons we might want to do this.

The environment itself sometimes has an uncertainty; take for instance a cube farm. The walls are not as precise as permanent building walls, and restless denizens of the farm may push at these confines, causing them to move slightly from day to day. When navigating from cubical walls, we may want to increase the minimum azimuth and lateral uncertainty limits to reflect this fact.

How much we reduce the uncertainty as the result of a correction is part of the art of fuzzy navigation. Some of the rules for decreasing uncertainty are:

1. After a correction, uncertainty must not be reduced below the magnitude of the correction. If the correction was a mistake, the next cycle must be capable of correcting it (at least partially).

2. After a correction, uncertainty must never be reduced below the value of the untaken implied correction (the full implied correction minus the correction taken). This is the amount of error calculated to be remaining.

Uncertainty may be manipulated in other ways, and these will be discussed in upcoming chapters. It should be apparent that we have selected the simple example of column navigation in order to present the principles involved. In reality, features may be treated discretely, or they may be extracted from onboard maps by the robot. In any event, the principles of fuzzy navigation remain the same. Finally, uncertainty may be manipulated and used in other ways, and these will be discussed in upcoming chapters.

Sensors, Navigation Agents and Arbitration

Different environments offer different types of navigation features, and it is important that a robot be able to adapt to these differences on the fly. There is no reason why a robot cannot switch seamlessly from satellite navigation to lidar navigation, or to sonar navigation, or even use multiple methods simultaneously. Even a single-sensor system may be used in multiple ways to look for different types of features at the same time.

Sensor types

Various sensor systems offer their own unique strengths and weaknesses. There is no "one size fits all" sensor system. It will therefore be useful to take a very brief look at some of the most prevalent technologies. Selecting the proper sensor systems for a robot, positioning and orienting them appropriately, and processing the data they return are all elements of the art of autonomous robot design.

Physical paths

Physical paths are the oldest of these technologies. Many methods have been used to provide physical path following navigation. Although this method is not usually thought of as true navigation, some attempts have been made to integrate physical path following as a true navigation technique. The concept was to provide a physical path that the robot could "ride" through a feature poor area. If the robot knew the geometry of the path, it could correct its odometry.

Commercial AGVs, *automatic guided vehicles*, have used almost every conceivable form of contact and noncontact path, including rails, grooves in the floor, buried wires, visible stripes, invisible chemical stripes, and even rows of magnets.

The advantages to physical paths are their technical simplicity and their relative robustness, especially in feature-challenged environments. The sensor system itself is usually quite inexpensive, but installing and maintaining the path often is not. Other disadvantages to physical paths include the lack of flexibility and in some cases cosmetic issues associated with the path.

A less obvious disadvantage for physical paths is one that would appear to be a strength: their precise repeatability. This characteristic is particularly obvious in upscale environments like offices, where the vehicle's wheels quickly wear a pattern in the floor. Truly autonomous vehicles will tend to leave a wider and less obvious wear footprint, and can even be programmed to intentionally vary their lateral tracking.

Sonar

Most larger indoor robots have some kind of sonar system as part of their collision avoidance capability. If this is the case, then any navigational information such a system can provide is a freebee. Most office environments can be navigated exclusively by sonar, but the short range of sonar sensors and their tendency to be *specularly reflected* means that they can only provide reliable navigation data over limited angles and distances.

There are two prevalent sonar technologies: Piezo-electric and electrostatic. Piezo-electric transducers are based on crystal compounds that flex when a current passes through them and which in turn produce a current when flexed. Electrostatic transducers are capacitive in nature. Both types require a significant AC drive signal (from 50 to 350 volts). Electrostatic transducers also require a high-voltage bias during detection (several hundred volts DC). This bias is often generated by rectifying the drive signal.

Both of these technologies come in different beam patterns, but the widest beam widths are available only with Piezo-electric transducers. Starting in the 1980s the Polaroid Corporation made available the narrow beam Piezo-electric transducer and ranging module that they had developed for the company's cameras. The cost of this combination was very modest, and it quickly gained popularity among robot designers.

The original Polaroid transducer had a beam pattern that was roughly 15 degrees wide and conical in cross-section. As a result of the low cost and easy availability, a great number of researchers adopted this transducer for their navigation projects. To many of these researchers, the configuration in which such sensors should be deployed seemed obvious. Since each transducer covered 15 degrees, a total of 24 transducers would provide the ability to map the entire environment 360 degrees around the robot.

The obvious way is usually a trap!

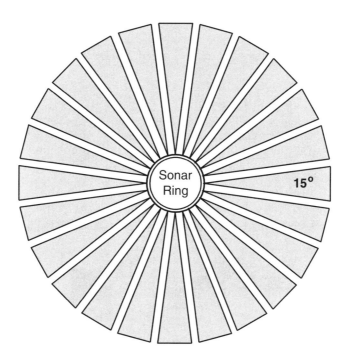

Figure 12.1. The ubiquitous "Polaroid ring" sonar configuration

The Polaroid "ring" configuration ignored the principles of proper sensor deployment.

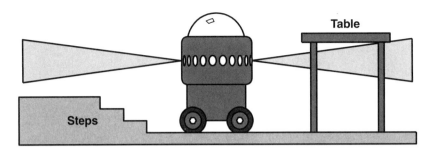

Figure 12.2. Vertical coverage of a narrow-beam ring

The most obvious weakness of the narrow-beam ring configuration is that it offers little useful information for collision avoidance as shown in Figure 12.2. This means that the cost of the ring must be justified entirely by the navigation data it returns.

From a navigation standpoint, the narrow beam ring configuration represents a classic case of static thinking. The reasoning was that the robot could simply remain stationary, activate the ring, and instantly receive a map of the environment around it. Unfortunately, this is far from the case. Sonar, in fact, only returns valid ranges in two ways: normal reflection and retroreflection.

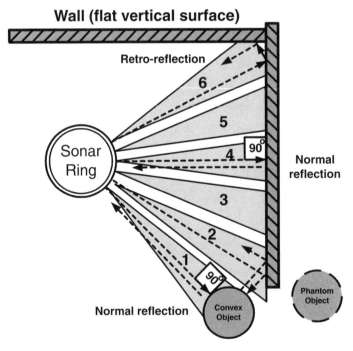

Figure 12.3 Modes of sonar reflection

Figure 12.3 shows two cases of normal reflection, one from a flat surface and one from a convex object. Also shown is a case of retroreflection from a corner. Retroreflection can occur for a simple corner (two surfaces) or a corner-cube. In the case of a simple corner such as in Figure 12.3, the beam will only be returned if the incidence angle in the vertical plane is approximately 90 degrees.

Unless the wall surface contains irregularities larger than one-half the wavelength of the sonar frequency (typically 1 to 3 cm), sensors 2, 3, and 5 will "go specular" and not return ranges from it. Interestingly, beam 2 returns a phantom image of the post due to multiple reflections.

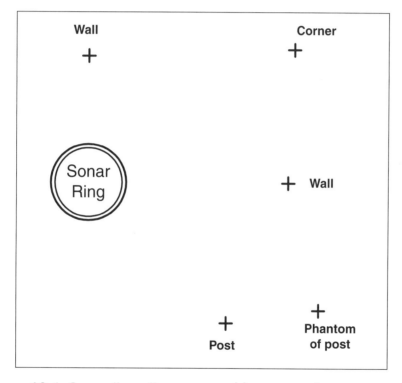

Figure 12.4. Sonar "map" as returned by narrow-beam sonar ring

Figure 12.4 approximates the sonar data returned from our hypothetical environment in Figure 12.3. If the data returned by lidar (Figure 10.2) seemed sparse, then the data from sonar would appear to be practically nonexistent.

Flashback...

I remember visiting the very impressive web site of a university researcher working with a sonar ring configuration. The site contained video clips of the robot navigating university hallways and showed the sonar returns along the walls ahead. The robot was successfully imaging and mapping oblique wall surfaces out to more than three meters.

The signal processing and navigation software were very well done, and it appeared that the configuration had somehow overcome the specular reflection problems of sonar. I was very puzzled as to how this could possibly be and found myself doubting my own experience.

Some months later, I was visiting the university and received an invitation to watch the robot in action. As soon as I exited the stairwell into the hall, the reason for the apparent suspension of the laws of nature became crystal clear. The walls of the hall were covered with a decorative concrete facade that had half-inch wide grooves running vertically on three-inch centers. These grooves formed perfect corner reflectors for the sonar. This is yet another example of a robot being developed to work well in one environment, but not being able to easily transition to other environments.

It would appear from Figure 12.4 that sonar navigation would be nearly impossible given so little data. Nothing could be further from the truth. The fact is that the two wall reflections are very dependable and useful. Since paths will normally run parallel to walls, we might be tempted to orient two or more sensors parallel on each side of the robot to obtain a heading fix from a wall. This would be another case of static thinking, and is far from the best solution. For one thing, with only two readings it would be almost impossible to filter out objects that did not represent the wall.

On the other hand, if the robot moves roughly parallel to one of these walls recording the ranges from normal reflections of a single beam, it can quickly obtain a very nice image of a section of the wall. This image can be used to correct the robot's heading estimate and its lateral position estimate, and can contain the number of points required to assure that it is really the wall.

The first bad assumption of the ring configuration is, therefore, that data in all directions is of equal value. In fact, data from the sides forward is usually of the most potential navigational importance, followed by data from the front. Since sonar echoes take roughly 2 ms per foot of range, and since transducers can interfere with each other, it is important that time not be wasted pinging away in directions where there is little likelihood of receiving meaningful data. Worse, as we will see, these useless pings will contaminate the environment and deteriorate the operation of the system.

The second bad assumption of the ring configuration is that the narrow-beam of the transducer provides the angular resolution for the range data returned. This may be true, but 15 degrees is far too wide to achieve reasonable spatial resolution.

The robot in Figure 12.4 can, in fact, determine a very precise vector angle for the reflection from the wall. The trick is for the robot to have a priori knowledge of the position and azimuth of the wall. If paths are programmed parallel to walls, then the robot can imply this from the programmed azimuth of the path. The navigation program therefore knows that the echo azimuth is 90 degrees to the wall in real space.

This is not to imply that the Polaroid ranger itself is not a useful tool for robot collision avoidance and navigation. Configurations of these rangers that focus on the cardinal directions of the robot have been used very successfully by academic researchers and commercial manufacturers of indoor mobile robots. Polaroid has also introduced other transducers and kits that have been of great value to robotics developers.

For Cybermotion robots, on the other hand, we selected wide-beam Piezo-electric transducers. Our reasoning in this selection was that the resolution would be provided by a priori knowledge, and that the wider beam width allowed us to image a larger volume of space per ping. The signal of a sonar system is reduced with the square of the pattern radius, so the 60-degree beam pattern we selected returned $1/16^{th}$ the energy of the 15-degree transducers. To overcome the lower signal-to-noise rations resulting from wide-beam widths, we used extensive digital signal processing.

A final warning about the use of sonar is warranted here. In sonically hard environments, pings from transducers can bounce around the natural corner reflectors formed by the walls, floor, and ceiling and then return to the robot long after they were expected. Additionally, pings from other robots can arrive any time these robots are nearby. The signal processing of the sonar system must be capable of identifying and ignoring these signals. If this is not done, the robot will become extremely paranoid, even stopping violently for phantoms it detects. Because of the more rapid energy dispersal of wide-beam transducers, they are less susceptible to this problem.

Thus, the strengths of sonar are its low cost, ability to work without illumination, low power, and usefulness for collision avoidance. The weaknesses of sonar are its slow time of flight, tendency toward specular reflection, and susceptibility to interference. Interference problems can be effectively eliminated by proper signal processing, and sonar can be an effective tool for navigation in indoor and a few outdoor applications. Sonar's relatively slow time of flight, however, means that it is of limited appeal for vehicles operating above 6 to 10 mph.

Fixed light beams

The availability of affordable lidar has reduced the interest in fixed light beam systems, but they can still play a part in low-cost systems, or as adjuncts to collision avoidance systems in robots. Very inexpensive fixed-beam ranging modules are available that use triangulation to measure the range to most ordinary surfaces out to a meter or more. Other fixed-beam systems do not return range information, but can be used to detect reflective targets to ten meters or more.

Lidar

Lidar is one of the most popular sensors available for robot navigation. The most popular lidar systems are planar, using a rotating mirror to scan the beam from a solid-state laser over an arc of typically 180 degrees or more. Lidar systems with nodding mirrors have been developed that can image three-dimensional space, but their cost, refresh rate, and reliability have not yet reached the point of making them popular for most robot designs.

The biggest advantage of lidar systems is that they can detect most passive objects over a wide sweep angle to 10 meters or more and retroreflective targets to more than 100 meters. The effective range is even better to more highly reflective objects. The refresh (sweep) rate is usually between 1 and 100 Hz and there is a trade-off between angular resolution and speed. Highest resolutions are usually obtained at sweep rates of 3 to 10 Hz. This relatively slow sweep rate places some limitations on this sensor for high-speed vehicles. Even at low speed (less than 4.8 km ph/3 mph), the range readings from a lidar should be compensated for the movement of the vehicle and not taken as having occurred at a single instant.

Figure 12.5. Sick LMS lidar
(Courtesy of Sick Corp.)

The disadvantages of lidar systems are their cost, significant power consumption (typically over 10 watts), planar detection patterns, and mechanical complexity. Early lidar systems were prone to damage from shock and vibration; however, more recent designs like the Sick LMS appear to have largely overcome these issues. Early designs also tended to use ranging techniques that yielded poor range-resolution at longer ranges. This limitation is also far less prevalent in newer designs.

Radar imaging

Radar imaging systems based on synthetic beam steering have been developed in recent years. These systems have the advantage of extremely fast refresh of data over a significant volume. They are also less susceptible to some of the weather conditions that affect lidar, but exhibit lower resolution. Radar imaging also exhibits specular reflections as well as insensitivity to certain materials. Radar can be useful for both collision avoidance and feature navigation, but not as the sole sensor for either.

Video

Video processing has made enormous strides in recent years. Because of the popularity of digital video as a component of personal computers, the cost of video digitizers and cameras has been reduced enormously.

Video systems have the advantages of low cost and modest power consumption. They also have the advantage of doubling as remote viewers. The biggest disadvantages of video systems are that the environment must be illuminated and they offer no inherent range data. Some range data can be extrapolated by triangulation using stereoscopic or triscopic images, or by techniques based on a priori knowledge of features.

The computational overhead for deriving range data from a 2D camera image depends on the features to be detected. Certain surfaces, like flat mono-colored walls can only be mapped for range if illuminated by a structured light. On the other hand, road following has been demonstrated from single cameras.

GPS

The Global Positioning System (GPS) is a network of 24 satellites that orbit the earth twice a day. With a simple and inexpensive receiver, an outdoor robot can be provided with an enormously powerful navigation tool. A GPS receiver provides continuous position estimates based on triangulation of the signals it is able to receive. If it can receive three or more signals, the receiver can provide latitude and longitude. If it can receive four or more signals, it can provide altitude as well. Some services are even available that will download maps for the area in which a receiver is located.

The biggest disadvantage to GPS is the low resolution. A straight GPS receiver can only provide a position fix to within about 15 meters (50 feet). In the past, this was improved upon by placing a receiver in a known location and sending position corrections to the mobile unit. This technique was called *differential GPS*. Today there is a service called WAAS or *Wide Area Augmentation System*. A WAAS equipped receiver automatically receives corrections for various areas from a geostationary satellite. Even so, the accuracy of differential GPS or a WAAS equipped GPS is 1 to 3 meters. At this writing, WAAS is only available in North America.

The best accuracy of a WAAS GPS is not adequate to produce acceptable lane tracking on a highway. For this reason, it is best used with other techniques that can minimize the error locally.

Thus, the advantages of the GPS include its low cost, and its output of absolute position. The biggest disadvantages are its accuracy and the fact that it will not work in most indoor environments. In fact, overhead structures, tunnels, weather, and even heavy overhead vegetation can interfere with GPS signals.

Guidelines for selecting and deploying navigation and collision avoidance sensors

The preceding discussion of various sensor technologies has provided some insight into the most effective use of navigation sensors. Following is a summary of some useful guidelines:

1. Try to select sensors that can be used for multiple functions. The SR-3 Security robot, for example, uses sonar for collision avoidance, navigation, camera direction, and even short-range intrusion detection.

2. Determine the sources of false data that the sensor will tend to generate in the environments where the robot will be used, and develop a strategy ahead of time to deal with it.

3. For battery-operated robots, power consumption is a prime concern. The average power consumed by the drive motor(s) over time can easily be exceeded by the power draw of sensors and computers. If this is allowed to happen, the robot will exhibit a poorer availability and a smaller return on investment to its user.

4. Deploy sensors to face in the directions where useful data is richest. Generally, this is from the sides forward.

5. Allow the maximum flexibility possible over the data acquisition process and signal-processing process. Avoid fixed thresholds and filter parameters.

6. Never discard useful information. For example, if the data from a sonar transducer is turned into a range reading in hardware, it will be difficult or impossible to apply additional data filters or adaptive behaviors later.

7. Consider how sensor systems will interact with those of other robots.

8. Sensor data processing should include fail-safe tests that assure that the robot will not rely upon the system if it is not operating properly.

9. Read sensor specifications very carefully before committing to one. A sensor's manufacturer may state it has resolution up to x and ranges out to y with update rates as fast as z, but it may not be possible to have any two of these at the same time.

Since Polaroid transducers have been the most popular sonar sensors for mobile robots, it is instructive to look at some of the configurations into which they have been deployed. Figure 12.6 shows a configuration that resulted from the recognition of the importance of focusing on the forward-direction. This configuration has sacrificed the rear-looking elements of the ring configuration in favor of a second-level oriented forward.

While this configuration provides better coverage in a collision avoidance role, it still suffers from gaps in coverage. The most glaring gap is that between the bottom beam and ground level. There is also a gap between the beams as seen in the top view. Since the beams are conical in shape, the gaps shown between beams is only representative of coverage at the level of the center of the transducers. The gap is worse at every other level.

While it would appear that the gaps in the top view are minimal, it should be remembered that a smooth surface like a wall will only show up at its perpendicularity to the beams. In the patterns of Figure 12.6, there are 12 opportunities for a wall to be within the range of the transducers and not be seen. For this reason, when two half-rings are used, the lower set of transducers is usually staggered in azimuth by one-half the beam-width angle of the devices.

Notice that this configuration has 26 transducers to fire and read. If the range of the transducers was set to 3 meters (10 feet), then the round-trip ping-time for each transducer would be about 20 ms. If no time gap was used between transducer firings, then the time to fire the whole ring would be just over one half a second. This slow acquisition rate would severely limit the fastest speed at which the robot could safely operate. For this reason, ring configurations are usually setup so that beam firings overlap in time. This is done by overlapping the firing of beams that are at a significant angle away from each other. Unfortunately, even with this precaution the process greatly increases the number of false range readings that result from cross beam interference.

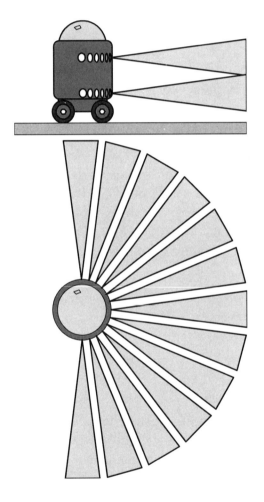

Figure 12.6. Dual half-ring configuration

Figure 12.7 demonstrates yet another variation that attempts to improve on the weaknesses of the ring. In this staggered ring configuration, transducers are mounted in a zigzag pattern so that their beam patterns close the gaps that plague the earlier examples. This same pattern can be duplicated at two levels as was done with the straight rings in the previous configuration. This configuration provides better coverage in the plan view, but at the expense of having even more transducers to fire.

Staggered configurations may increase the number of transducers per a half-ring from 13 to 16 (as in Figure 12.7) or even to 25, depending on the overlap desired. By this point, we can see that getting good collision avoidance coverage with a ring configuration is a bit like trying to nail jelly to a tree. Everything we do makes the problem worse somewhere else.

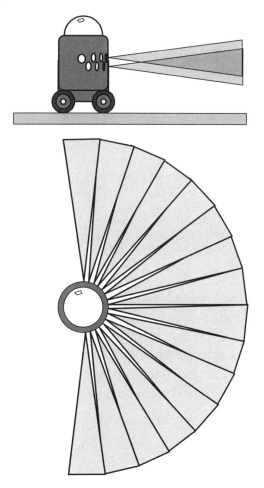

Figure 12.7. Staggered-beam configuration

Several commercial robots have achieved reasonable results with narrow-beam transducers by abandoning the ring configuration altogether. Figure 12.8 is representative of these configurations, which concentrated the coverage in the forward path of the robot. This concentration makes sense since this is the zone from which most obstacles will be approached.

The weakness of the forward concentration pattern is that a robot can maneuver in such a way as to approach an obstacle without seeing it. The solid arrow in Figure 12.8 indicates one type of approach, while the dashed arrow shows the corresponding apparent approach of the obstacle from the robot's perspective.

To prevent this scenario, the robot must be restricted from making arcing movements or other sensors can be deployed to cover these zones. The commercial robots that have used this approach have typically added additional sensors to avoid the problem.

At Cybermotion, we developed our strategy in much the same way others did. We took our best guess at a good configuration and then let it teach us what we needed to change or add. Unlike most of the industry, we decided to use wide-beam Piezo-electric transducers, and to use extensive signal processing. We never had reason to regret this decision.

All sonar transducers have a minimum range they can measure. While this minimum range is usually less than .15 meters (half a foot) for electrostatic transducers, it is typically as much as .3 meters (1 foot) for Piezo-electric transducers. In order for the SR-3 to navigate from walls close to the sides of the robot, the side transducers were recessed in slots in the sides of the robot so that it could measure distances as low as .15 meters.

Early configurations had two forward transducers and two side transducers only. The forward transducers were used for collision avoidance and to correct longitudinal position. The beam patterns of these transducers were set to be cross-eyed in order to provide double coverage in front of the robot. Because most objects in front of the robot can be detected by both beams, the robot can triangulate on objects. In fact, the SR-3 uses this technique to find and precisely mate with its charging plug.

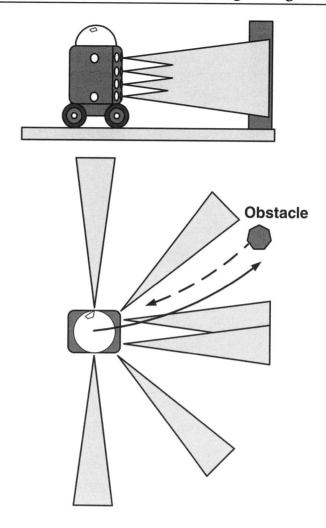

Figure 12.8. Forward concentration strategy

The two forward beams were also directed downward so that they illuminated the floor a bit less than a meter in front of the robot. As seen in Figure 12.9, objects lying on the floor tend to form corner reflectors with that surface, making them easily detectable by retroreflection as shown in the side view. By programming the robot to stop at distances greater than the distance from the transducer to the floor (S), it can be made to stop for relatively small obstacles.

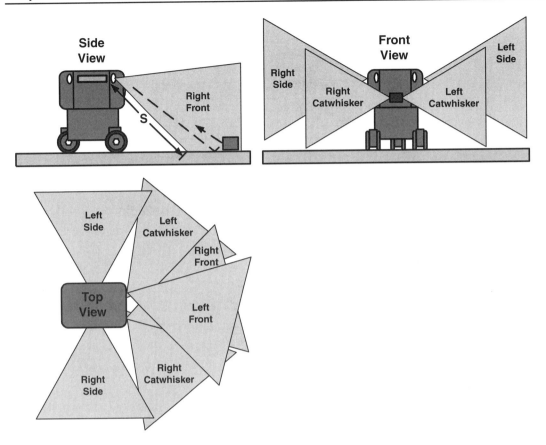

Figure 12.9. Wide-beam configuration with cat whiskers

The signal processing can distinguish between reflections from carpet pile and obstacles. Although this configuration worked extremely well in straight forward motion, it suffered from the same vulnerability as demonstrated in Figure 12.8 when steering in a wide arc.

To avoid the possibility of approaching an obstacle in this way, the configuration was enhanced with two transducers that came to be known as "cat whiskers." The final configuration for the SR-3 is shown in Figure 12.9.

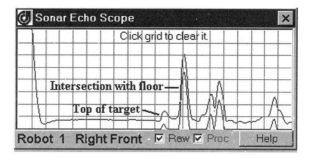

Figure 12.10. Wide-beam retroreflection from corner of target and floor[1]
(Courtesy of Cybermotion, Inc.)

Lidar has occasionally been used in a downward mode to illuminate an arc in front of the robot. Configurations like that shown in Figure 12.11 have been used by several robots, and have even been added to Cybermotion robots by third parties.

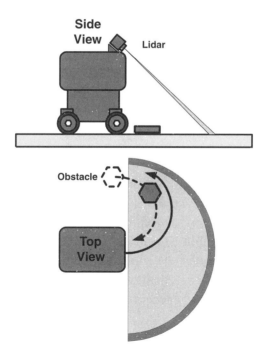

Figure 12.11. Downward-looking lidar for collision avoidance

---

[1] The upper trace is the raw sonar echo data and the lower trace is the same data after signal processing.

A downward-looking lidar has a vulnerability to sweeping over low-lying obstacles during tight turns as shown in Figure 12.11. Downward-looking lidar is also of very limited use in navigation, and suffers from specular reflection on shiny surfaces. This can make the system believe it has detected a hole.

Downward-looking lidar is one of the few sensors that can detect holes and drop-offs at sufficient range to allow the robot to stop safely. Given the high cost and power consumption of the lidar, and its limited use for navigation, this configuration is difficult to justify based primarily on hole detection. In permanently installed systems, dangers such as holes and drop-offs can simply be mapped ahead of time.

An alternative and perhaps more economical configuration of a downward-looking sensor has been used by the Helpmate robot. This configuration consists of simple beam-spreaders and a video camera. The Helpmate configuration has the advantage of projecting several parallel stripes, and thus being less vulnerable to the turning problem.

The purpose here, as in other chapters, is not to recommend a configuration. The goal is only to point out some of the considerations associated with selecting and deploying sensors so as to achieve dependable collision avoidance and navigation at the lowest cost in money and power. New and more powerful sensors will be available in the future, but they too will have vulnerabilities that must be taken into consideration.

Navigation agents

A *navigation agent* is simply a routine that processes a certain kind of sensor information looking for a certain type of navigation feature. There may be many types of *navigation agents* in a program and at any given time there may be one or more instances of a given *agent* running. It should be immediately apparent that this is exactly the kind of problem that the concept of *structured programming* was developed to facilitate. Because the term *navigation object* sounds like something in the environment, the term *agent* has become relatively accepted in its place.

Agent properties and methods

Like all structured programming objects, navigation agents will have properties and methods[2]. The main method of a navigation agent is to provide a correction for the

---

[2] Refer to Chapter 2 for a brief description of the concepts behind structured programming.

robot's position and/or heading estimate as discussed in the previous chapter. Some agents will be capable of updating all axes, while others will be able to correct only one or two.

Simply because an agent can offer a correction to two or more axes does not mean that these corrections should always be accepted. It is quite possible for an agent to provide a relatively high Q correction for one axis while offering an unacceptably low Q for another axis. In the previous chapter, we saw that the quality of a y correction from the column in Figure 10.11 was lower than that for its x correction. This was a result of the observational uncertainty induced by heading uncertainty. This brings us to the cardinal rule about agents.

Navigation agents must never correct the position estimate themselves, but rather report any implied correction to an arbitrator along with details about the quality of the correction.

There is another potential problem associated with multiple agents. If one agent reports a correction that is accepted, other agents must have a way of knowing that the position or heading estimate has been corrected. If an agent is unaware that a correction was made while it was processing its own correction, the data it returns will be in error. There are two reasons for this: first, the data collected may be based on two different reference systems; and second, even if this is prevented the same basic correction might be made twice.

There are several ways to contend with the problem of other agents causing the position estimate to be updated. One method is to provide a globally accessible *fix count* for each axis. When an axis is updated, that counter is incremented. The counters can roll over to prevent exceeding the bounds for their variable type. As an agent collects data to use for correcting the position estimate, it can read these counters to determine if a correction has been made that could invalidate its data.

If an agent has been preempted in this way, either it can dump its data, or it can convert all its existing data by the amount of the preempting correction. A word of warning about converting data is in order—if data is converted repeatedly, rounding and digital integration may begin to degrade the collection significantly.

Arbitration and competition among agents

The code that activates, deactivates, and takes reports from agents is the arbitrator. The arbitrator may activate an agent as the result of references to navigation features as specified in a program, or as the result of an automatic behavior such as map interpretation. The arbitrator must keep track of the agents running, and assure that they are terminated if no longer appropriate.

Since agents are essentially tasks in a multitasking environment, it is essential to assure that they receive machine resources when needed. The most efficient way to assure this is to activate an agent's thread whenever new data is available from its sensor. If the data processing routine acts on a large amount of data at a time, it may be prudent to release the thread periodically to allow other tasks to run.

When an agent reports a potential correction to the arbitrator, the arbitrator must decide whether to accept it, decline it, or suspend it. This decision will be based on the image Q (quality) of the data and its believability Q. If the correction is suspended, then the arbitrator may wait some short period for better data from another agent before deciding to partially accept the marginal correction.

The arbitrator will accept a correction proportional to its Q. For example, if a lateral correction of 0.1 meters is implied, and the fix Q is .25, the arbitrator will make a correction to the lateral position estimate by .025 meters. This will be translated into a combination of x and y corrections.

Who to believe

Since agents can interfere with each other, it is up to the arbitrator to assure the corrections that are most needed are accepted. Here we have another use for the robot's uncertainty estimate. For example, it is quite possible for a robot to have an excellent longitudinal estimate and an abysmal lateral estimate or vice-versa. The uncertainty for these axes will make it possible to determine that this situation exists. If one agent is allowed to repeatedly correct one axis, then it may interfere with a correction of the other axis.

Take, for instance, a robot approaching a wall with sonar. Assume that there are two agents active; one agent is looking for the wall ahead in hopes of providing a longitudinal correction, while the other agent is collecting data from a wall parallel to the robot's path, trying to obtain lateral and heading corrections. The first agent can offer a correction every time its forward sonar transducers measure the wall ahead, but the second agent requires the vehicle to move several feet collecting sonar ranges normal to the wall. If the first agent is allowed to continually update the longitudinal estimate, it may repeatedly invalidate the data being collected by the second agent.

If a correction that has a high repetition rate is allowed to continually correct one axis, then it may drown out any corrections for another axis. It is therefore up to the arbitrator to determine if this is potentially the case, and to prevent it from happening.

Changing one's mind

At this point, we will address one remaining consideration. No matter how careful we are in filtering out bad data, there will always be some that gets through. For example, the LMS lidar returns range fields that discriminate between normal objects and retroreflectors. As a feature sensor, this mode of the LMS is vastly more discriminating than most other types of sensors. Normally, the only false signals from this sensor come from unexpected retroreflectors such as the taillight assemblies of vehicles and the safety tape on forklifts.

However, surprisingly the cellophane wrap used on some pallets can duplicate the retroreflector affect by forming tiny corners in its folds. Unfortunately, this wrap is often found in close proximity to the posts where reflectors may exist in warehouses. If this happens near an expected target location, and the true target is masked, it can cause an error so subtle that the fuzzy logic may partially accept it.

It is important to remember that false navigation data can actually imply a correction that is closer to our expected position than a valid correction would have implied! It is therefore essential that the navigation filtering process never close down to the point that it cannot correct its last decision if that decision should turn out to be false.

It is theoretically possible that the arbitrator could maintain more than one position model for the robot for short periods. That is, different agents might be allowed to correct different models, only one of which was selected by the arbitrator to show to the outside world. In these cases the arbitrator would maintain the ability to revert to an alternate model. I have never attempted this.

Summary

Before any serious attempt is made at writing navigation routines, it is important to set up a structure under which they will all operate. By defining this structure, we can assure that not only will our original navigation process be well defined and orderly, but also it will be a simple matter to add agents later. Finally, it should be pointed out that we have again seen the central role that is played by the uncertainty estimate in the navigation process. This is not the last use we will find for this vital calculation.

Instilling Pain, Fear and Confidence

In earlier chapters, we have seen parallels between the ways animals, people, and robots cope with problems such as action planning and navigation. At this point, it is useful to consider the purpose of certain basic emotions in survival. To a large extent, autonomous robots will be subject to the stresses analogous to those experienced by living creatures in that they both must move about in changing environments and accomplish certain basic tasks. Whether we are prone to the anthropomorphic or not, robots will need methods to cope with these same sorts of situations.

Pain and annoyance

Few of us enjoy pain, but as a behavior modifier, it plays a critical part in our survival. For example, one of the most devastating side effects of diabetes is that it reduces circulation and sensation in the extremities. Without pain, a diabetic may have no idea that she has gotten a pebble in her shoe. The results can be devastating.

Pain implies a potentially damaging event, while annoyance implies a situation that simply impedes progress. The difference in response to pain and annoyance is mostly a matter of degree, so for simplicity we will consider annoyance to be low-amplitude pain. We will also lump threatening events (near misses) into this category.

Virtual pain

To imbue a robot with a true sense of pain is of course a bit beyond the reach of the current state-of-the-art in software. It is easily within our reach, however, to observe the self-protective reactions animals and people show to pain, and to devise ways to achieve similar behavior modification in robots. There are two fundamental responses to pain:

1. Pain provides immediate negative feedback and temporary behavior modification we will call reduced *confidence*.

2. Pain modifies the way we do things in such a manner as to avoid known sources of pain in the future. The anticipation of pain is a sense of *danger*, and this anticipation induces a loss of *confidence*.

For example, if we get a pebble in our shoe while walking, until we can remove it we will immediately slow our pace, and put less weight on the affected foot. In the long term, if this happens every time we walk along a certain gravel footpath we may find ourselves walking at the edges of the path, or avoiding it altogether. We have associated this path with a mild danger of pain.

We have only to decide what conditions we wish to avoid or minimize, and what response to elicit if they occur. Bumper strikes are an easy first choice. Even though our robot will almost certainly have some form of noncontact collision avoidance such as sonar, it should also have a mechanical bumper as a final safety device. If the bumper is activated the robot will be disabled from any further motion, at least for some time, in the direction it was moving at contact. The bumper strike may have been caused by something outside of the robot's control, such as someone bumping into the robot, but it happened just the same.

Flashback...

I am reminded of a funny example. We were demonstrating our SR-3 Security robot at a huge security trade show. In the early years, we had dutifully obeyed the show rules book concerning moving equipment and had kept our robot inside a fenced demonstration area. Visitors would naturally ask us why the robot had to be fenced off if it was designed to operate safely among people. So, by the time of this particular show we had gotten in the habit of turning our robot loose in the aisles under the maxim that "It is easier to get forgiveness than it is to get permission."

On this particular occasion I was watching the robot coming down the aisle toward our booth and saw that there was a heavy-set lady standing in front of a neighboring booth with her back to the aisle. She was apparently bent on making some very vocal point with the defenseless vendors in the booth and seemed oblivious to their warning gestures at the approaching robot. As she made her closing arguments, she was backing directly into the path of the robot, her arms flailing.

As the robot came close to the woman, it calculated that they were on a collision course and pulled to a complete stop. Unfortunately, she did not. Arms still flailing, she backed into the robot, almost tripping over the bumper. Her reaction was immediate and merciless. She spun around, and undaunted by the mechanical nature of her adversary she immediately began unleashing a verbal tirade upon the hapless robot. This assault, combined with the joy of their own reprieve, caused the vendors in the booth to lose all control and begin laughing convulsively. The derision from her previous victims only served to amplify the intensity of the woman's attack, and the encounter ended with a well-placed blow from her purse and a promise to report the offensive beast to show management[1].

There is a lesson here. Bad things will happen to good robots and robots are presumed guilty unless proven innocent. If there is ever an accident involving humans and robots, the robot is going to be blamed unless its keepers can provide absolute documented proof that the incident was caused completely by the human.

When a robot resumes operation following a bumper strike or other *pain event*, it should do so under *reduced confidence*. We will consider *confidence* shortly, but for now let's consider simple ways we can teach our robot to avoid bad things. Paths or areas where bumper strikes and other unpleasant things occur should ideally be remembered and avoided. If they cannot be avoided, then they should be navigated with more caution.

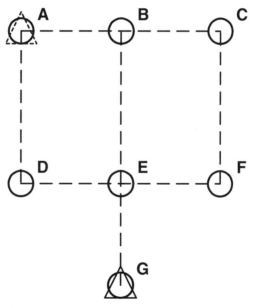

Figure 13.1. Penalizing path segments

---

[1] She did report the robot, but within an hour it had mysteriously escaped incarceration in our booth and was venturing out again and without further incident.

Avoiding pain and humiliation

If the robot is operating from scripted path programs, then these programs will consist of short paths that will be concatenated by a planner to take the robot from one place to another at the lowest possible *cost*[2].

Figure 13.1 shows a simple case of path segment navigation. For example, the path between G and E is described in a program script we will call path GE. It is clear that the robot can get from its current position at *node* G to *node* A most quickly by concatenating paths GE, ED, and DA, or by concatenating GE, EB, and BA. It could also use a more circuitous route of GE, EF, FC, CB, and BA as an alternative if the shorter paths are unavailable.

Virtual danger

Normally the base *cost* for a given path segment is initially a simple function of the calculated time required to transit it. As the robot runs the path, the actual time that is required to transit it may be longer than calculated. For this reason, it is a good idea to modify the calculated transit cost by a weighted average of recent transit times. An easy way to do this without saving an array full of time values is:

$$\text{Transit\_Time} = ((1–TC) * \text{Transit\_Time}) + (TC * \text{New\_Time})$$

Here, TC is a number greater than zero but less than one. TC is the time constant factor for how aggressively the smoothing function tracks the newest transit time. For example, if TC is 0.1, then the transit time will be 10% of the latest value and 90% of the historical value. This method of weighted averaging is not only fast and efficient, but also it gives more importance to recent data. To this updated transit-time cost, we can add a *danger* cost.

$$\text{Path\_Cost} = \text{Transit\_Time} + \text{Danger\_cost}$$

Notice that the units here are arbitrary as we are merely looking for numbers to compare between paths. If we keep the Path_Cost units and the Danger_Cost units in the same time units as the Transit_Time, then we don't have to worry about translation factors and calculations run that much faster.

---

[2] See Chapter 8 for a more comprehensive description of path programming.

In the above example, there are two equally fast ways to get from G to A. For simplicity, we will assume that the transit time for each path is the same as the calculated time. If a *danger* factor is added to the transit cost to reflect things that have gone wrong on a path in the past, then this element can be increased with each unpleasant episode, and can be gradually diminished with the passing of time. This way, the robot will eventually go back to use the path again. Transiting the path without incident can also be used as an opportunity to decrement its *danger* factor by an additional amount.

Different types of painful or delaying experience may occur on the same path. Each type of unpleasantness will increment the path cost by an amount proportional to its seriousness. These weighting factors should be blackboard variables[3] so that they can be modified later for differences in environments.

Thus, if path DA is the scene of a bumper strike, then the slightest *fear* penalty to its *cost* will send the robot around the route GE-EB-BA. If the EB segment subsequently begins collecting danger factor, it would take more of a danger factor to force the robot around the long route than it took to force it from the first route because the transit cost of this alternate route is higher. This should be intuitively logical. In fact, the robot might go back to using the first path until and unless additional problems occurred on the DA segment.

On the other hand, consider a case where the robot had been penalized on the DA path and reverted to the GE-EB-BA route, and then experienced pain on the BA path. In this case, it would not consider the longer route as this would also have the BA danger factor cost. With nothing more than addition and subtraction, we have created a very effective protective response for the robot.

Sources of virtual pain

Some of the sources of virtual pain and annoyance that may indicate when a path should receive a *danger* penalty include:

1. Bumper strikes have occurred.

2. The robot has been forced to circumnavigate obstacles.

3. The robot has been recalled from the path due to blockages.

---

[3] See Chapter 6 on the importance of blackboards and of avoiding embedded constants.

4. The drive or steer servo has experienced current limiting or stalling.

5. Excessive numbers of navigation features have been missed.

6. Position uncertainty has grown excessive.

At some level of danger, it is not safe to use a path, even if it is the only route available to a destination. At this point, a path is considered to be suspended from service. Eventually a path must be restored to service or we will gradually find that there are fewer and fewer routes available to the robot. Paths *suspended* by problems should always return to service after some period.

In the CyberGuard system, we provided a method by which the operator could click on a path and either temporarily *suspend* it or permanently *disable* it. This feature proved to be very useful during extended periods of construction or moving. Remember also that there may be a mirror image path going the other direction. Generally, if this is the case then this mirror path should receive the same danger penalty.

Forgetting danger

To assure that the robot will eventually use a path again, it is necessary that the *danger factor* decay with time. If we decrease (decay) the penalty in a path too quickly over time we will repeatedly expose the vehicle to situations that may require operator intervention or even cause damage. If we wait too long, we may disable a useful path needlessly for hours due to a very transient problem.

One simple solution to this dilemma is to decrease the decay rate every time a path is taken out of service, and increase the decay rate a little whenever the path has run well. Both a minimum and maximum decay rate can be specified to keep the recovery time reasonable. In this way, recurring problems will cause the robot to avoid the path for longer periods.

The base decay rate for danger factors must be a blackboard variable as well. The reason for this is that different environments tend to exhibit different danger persistence times. For example, in an office building a security robot will encounter cleaning personnel and their equipment fairly frequently, especially early in the evening. On the other hand, these blockages do not stay in one place longer than a few minutes. In a warehouse, however, blockages encountered after hours are almost sure to remain in place all night.

An ideal system would take this concept to the next level by associating dangers with times of day, days of the week, and other conditions to modify danger values even more accurately. Even so, the very simple methods just discussed bring a great reward in system performance.

Danger objects and force fields

If force field goal seeking is the preferred approach to navigation, then we need only define a transient *danger object*. This object is similar to a fixed obstacle, but with a repelling force that is variable. The repelling force serves the same purpose as the *danger* cost of a path segment, and it should be incremented and decremented in much the same manner.

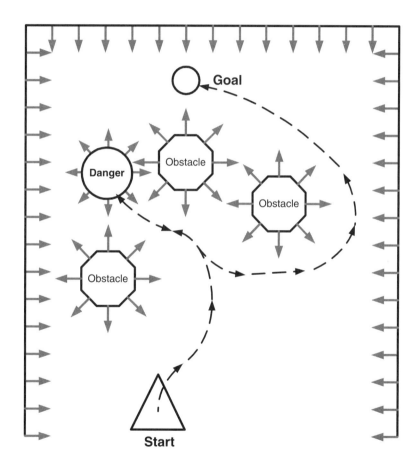

Figure 13.2. Placing danger objects on a force field map

197

In Figure 13.2, the robot has encountered a source of pain and has planned a new route to the goal. As long as the danger object's field strength remains significant, it will elevate the floor of any nearby valleys. Since route planning follows valleys in the force field, the more the floor of the valleys in the area of the danger become elevated, the more likely that the robot will avoid using them. As the force field decays, the robot will tend to retry the shorter routes. As in the case of weighted path segments, the greater the difference in route lengths, the sooner the robot will retry the shortcut.

There is a simple elegance to this method, in that the processing for the *danger object* is almost exactly the same as for other obstacles. Furthermore, if there is sufficient room around a danger object, very subtle path modifications can be made to precisely avoid it. In the virtual path method we knew only to avoid a whole path segment, but not precisely where the danger was on the path.

A hybrid map and path approach

An alternate approach to the problem is to use the path method, but to augment it with simple force field concepts. If this is done, then danger objects can be referenced in the properties of path segments, along with their coordinates. While the process significantly increases the computations required for path planning, it is still far less computationally expensive than the full-blown force field map approach.

A hybrid approach can maintain all the explicit instruction capability of the virtual path method, while inheriting some of the power of the force field method. The best method between these three alternatives will depend upon the application and its environment.

The purpose of virtual confidence

Generally, fear and caution are the antithesis of *confidence*, with a loss of *confidence* being a rational response to pain or danger. For mathematical reasons it is cleaner to simulate the positive emotion of *confidence* than its negative counterparts of fear and caution[4]. There are many reasons a person might lose *confidence*, but the four most useful to simulate are:

1. Aftermath of pain

2. Anticipation of danger

---

[4] Originally, we implemented these functions as separate reactions including one called "caution." Later, the unity of these concepts became apparent and we began consolidating them.

3. Disorientation

4. Diminished fitness

For example, after being in an accident we will tend to drive less aggressively for some period. Even after many years, if we find ourselves in the same location or situation in which the accident occurred, we may become less confident and drive more defensively. Finally, if we feel somehow incapacitated or realize we may have missed a turn, we will have a similar reaction.

Effects of virtual confidence

The proper responses of an autonomous robot to a loss of *confidence* are therefore relatively clear. While a robot remains in a state of diminished confidence, it should:

1. Reduce speed as a percentage of programmed speed

2. Reduce planning deceleration to begin slowing further away from obstacles

3. Reduce acceleration

4. Reduce torque limits on drive servos

By doing these things, we give the robot more time to sense stealthy obstacles, and we assure that if there is an incident it will be less severe than it would otherwise have been. It is important to remember that the impact force of our robot is proportional to its weight, but proportional to the square of its velocity! If we have these various control parameters in an accessible form, and not as constants[5], then modifying them should be a simple matter.

Going back to our example from Figure 13.1, let's assume that we have experienced problems on the DA path segment and thus stopped using the GE-ED-DA path in favor of the GE-EB-BA route. Now let's assume that we experience a problem along EB that makes this route more costly (risky) than the original GE-ED-DA route, but not as costly as going all the way around the GE-EF-FC-CB-BA route. If we had weighted our penalty points higher, then perhaps the long route would have been the best remaining approach, but assume this is not the case. We are going to go back to using the GE-ED-DA route, but we need to reflect the increased risk of this route by reducing the robot's confidence.

---

[5] See Chapter 6.

Calculating virtual confidence

If we normalize the value of *virtual confidence* to a number from zero to one, then we can simply multiply the current programmed speed by the confidence and immediately develop a slowing response to lost confidence. The same can be done for motor torque limits, acceleration, and deceleration. If we later find this to be a suboptimal reaction, we can easily add a simple trapezoidal function to the process as we have discussed for other processes.

Confidence and pain

Any given dangerous experience should erode confidence in proportion to its seriousness. For example, if the robot is forced to circumnavigate an unexpected obstacle, it might experience a 50% decrease in *confidence*. A bumper strike on the other hand might temporarily reduce the *confidence* to zero, and require an operator "*assist*" to bring it back to, say, 33%. An operator assist could be a local or remote restart command that indicates that the conditions are now safe. The details of this mechanism will necessarily depend on the application and the environment.

Once pain has caused the robot to become less confident, it should stay less confident until it has successfully driven a sufficient distance without further incident. This implies that if the robot became less *confident* due to navigation difficulties (high uncertainty) then it will need to acquire enough navigational references to reduce the uncertainty, and then drive a sufficient distance to assure it is out of the area of danger.

Anyone who has driven country roads late at night knows that if one deer crosses the road in front of you, there is a very good chance others are following and that it is wise to slow down. The same is true with most of the environmental problems a mobile robot will face. For example, if there is one trash can in the aisle, there will probably be more.

The amount of *confidence* loss induced by a pain or near-pain experience should therefore begin to erode with successful travel at a rate specified in a blackboard variable.

Confidence and danger

The danger of navigating a path segment is the sum of the known *inherent* danger and the *learned danger* we discussed earlier. All areas or paths have an inherent danger associated with them; the narrower the clearance between obstacles becomes, the higher the risk.

$$Danger = Inherent\_danger + Learned\_danger$$

In the case of the virtual path method of navigation, the risk must be implied from other parameters. One indicator can be the collision avoidance settings. Paths in narrow areas must use closer stop distances applied to the collision avoidance in order to execute.

Thus, if the forward stop distance is beyond some relatively high threshold, there is no inherent danger implied from it. If it is less than this value, then an inherent *danger confidence* factor is calculated proportionally. The same is done for the side-stop ranges, and these two factors are multiplied by each other. Therefore, the robot knows the *inherent danger* factor from these settings.

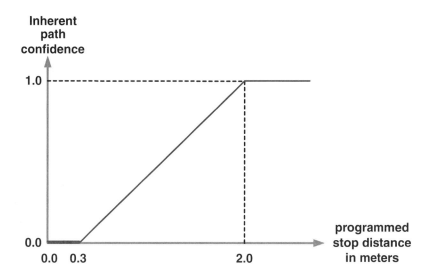

Figure 13.3. Determining inherent confidence from forward sonar stop distance

The curve in Figure 13.3 is that used by the SR-3 robot. Notice that when the forward stop distance setting drops below .3 meters, the robot has zero confidence and will not drive. A special instruction allows the robot to violate this range for docking maneuvers.

In the case of the force field model, the field level at the robot's position implies the inherent risk of traveling over this area. Instead of dealing with this component separately in the force field method, we can lump the calculation with that of position uncertainty. We will discuss this shortly.

Position confidence

Here we find yet another use of the robot's uncertainty estimate. If a robot is moving without good navigational feedback, then its position may be well out of the planned path or area. This can lead to serious consequences, particularly when there are mapped dangers that the robot cannot detect with its sensors. Such dangers might include drop-offs, overhead obstacles, and other moving machinery. The chances of avoiding these types of obstacles, and the potential damage that could be caused by an encounter with one of them are increased at higher speeds.

Conversely, slowing down gives the robot's sensors a better chance of imaging a navigational feature. If traffic is blocking the robot's view of navigation features, then not only will slowing down increase the chances of glimpsing some features through the traffic, but fundamentally it is a safe thing to do in a crowded situation.

We cannot simply slow the robot down in all areas because this would diminish the work it could accomplish in a given time, and thus the payback period. It is crucial that the robot run as fast as is safe whenever possible.

For example, assume our robot is operating near the entrance of a theater when the show ends. The crowd of people may well block the robot's view of its navigation features and disrupt navigation. If the robot continues to move rapidly, it will experience greater and greater navigational uncertainty and may become *unreferenced*. Worse yet, it might experience an exchange of kinetic energy with an unwary patron!

Uncertainty has a lateral, a longitudinal, an azimuth, and possibly a height component. A conservative method of evaluating the danger from uncertainty is to use the vector sum of the lateral and longitudinal components. A less conservative method is to simply use the lateral uncertainty since it is most likely to represent the threat of contact with a mapped or unmapped danger. But what should this value be compared against? If the robot is driving through a narrow aisle, the tolerable uncertainty is obviously less than if it is traveling across an open lobby.

In the virtual path approach of navigation, the allowable deviation from the path is usually set by an instruction. This can be explicit, or implied from a navigational reference to walls or aisles. During circumnavigation, the robot will not be permitted to move away from the path by more than these left- and right-hand limits minus its current lateral uncertainty. Likewise, during normal navigation the robot should not be allowed to continue moving if its uncertainty is greater in magnitude than the path limits. The smaller of the lateral path limits then makes an acceptable parameter against which to compare lateral uncertainty in calculating *inherent confidence*.

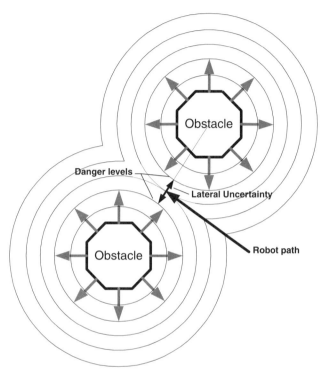

Figure 13.4. Deriving confidence from uncertainty in a force field system

Again, there is an equivalent way of deriving a *position confidence* factor when using force field routing. A reasonable approximation of the danger posed by position uncertainty is shown in Figure 13.4. In this case, we are simplifying the calculation by assuming that the lateral uncertainty is the most critical element. Since paths usually follow valleys in the force field, this is usually the case. In a few cases there may actually be a higher field level under the footprint of the robot's whole uncertainty, but the benefit for finding it is minimal. Again, we are dealing with nondeterministic values that need only elicit a generally appropriate response.

By simply calculating the floor level at a distance equal to the lateral uncertainty to the left and right, we find the danger (strength of the field) if we are out of position by the full magnitude of the uncertainty. The worst (highest) of these two values is used to determine the position confidence.

There are normally two thresholds used by the router in a force field map. The higher of these thresholds is the level at which the robot is prohibited from using the position. A lower threshold may also be specified. For areas with field strength below this lower threshold, the router is allowed to use the shortest path and not follow the valley floor.

These two levels nicely frame the position confidence. As shown in Figure 13.5, if the field level detected laterally in Figure 13.3 is at or below the lower level, then the *confidence* is 1.0 (100%). If the field level reaches the upper limit, the *position confidence* drops to 0%.

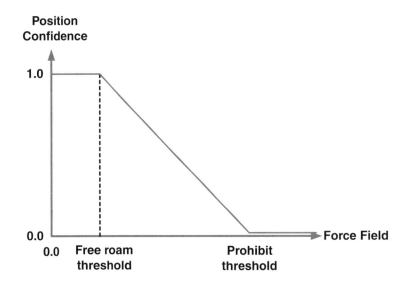

Figure 13.5. Determining position confidence from the force field level at the worst uncertainty

Since the force field level at any given point is a combination of the inherently known obstacles, and the danger objects the robot has learned, three elements of confidence are all derived at once. This is another example of the inherent elegance of force fields, but the calculations are far more complex than those for the virtual path model.

A weakness of the force field strategy is the lack of methods for preprogramming parameters along specific paths. Since there is not a path program in which to state the settings for a robot running under the force field method. Therefore, to operate at optimal safety levels the collision avoidance stop ranges must be derived from the field levels around the robot. Thus, the flow of information is the mirror of that for the virtual path method.

Confidence and virtual fitness

Here we add another anthropomorphic concept that is a metaphor for health and acuteness of the senses. For example, let's assume that a robot has very low batteries. If it attempts to accelerate at its full rate to reach its top speed, then the robot will probably experience a servo stall. Worse, servo performance may become sluggish during emergency maneuvers, and the robot will be wasting battery power it may need to conserve for the return to its charger.

In Cybermotion robots, the platform automatically quits accelerating as the servo begins to approach its current limit or full power command. Even so, we found it beneficial to reduce the velocity target as the batteries became depleted.

Another example of low fitness would be one or more sensor systems being partially or completely disabled. There are many reasons why this might happen. For example, in the presence of white noise sound sources like a leaking air line, sonar will be less effective than in a quiet environment.

In other cases, sensors must intentionally be desensitized to accomplish a maneuver. As mentioned earlier, it may be necessary for the robot to completely ignore a collision avoidance sensor system in order to complete an operation such as mating with its dock. If this is done, however, the robot's speed and power must be severely limited.

Flashback...

An interesting example of the effectiveness of such precautions occurred shortly after the story of the robot that tried to run me down because of a typo in one of its programs (see Chapter 2). Because of that experience, we had modified our vehicle code to slow the robot down as a result of several of these fitness factors. It was in fact an early form of *fitness confidence*.

I was visiting a Navy site that was using our robots. For some months, I had been monitoring reports on the effectiveness of our collision avoidance, and there had been several cases in which the robot had failed to stop for people in its way. However, in other tests it had successfully detected and stopped for very small objects on the floor. The path programs where these incidents had occurred had been checked, and did not contain errors such as the one that had resulted in the pushy robot I had experienced.

I was very puzzled by this. In any event, shortly after I had given the new programs to the Navy technician, I asked him if there were any problems. He told me that he had removed the new software because it was obviously defective. He said that the robot would barely move when it left its charger.

The mystery was solved! When the robot approached the charger to dock, it had to physically contact the charger. The program therefore disabled the sonar collision avoidance as the robot approached the charger. In the path program leading away from the charger, the programmer had placed an instruction to reenable the sonar collision avoidance. We had verified this beforehand.

The bug was introduced when someone had programmed a second path leading from the dock, but had not reenabled the sonar on that path. If the robot left by one route, all was well. If it left the charger by the other route, the robot would have disabled collision avoidance for some distance. In this case, it would eventually execute a path where the collision range setting was changed and it would suddenly become civil again. In the earlier version of the code, the robot was perfectly free to run about at full speed with its sonar system disabled if told to do so by its programmer.

We had serendipitously discovered an added advantage to calculating confidence from these settings; robots accidentally running with unsafe settings became very obvious!

At this point we have discussed how to calculate the five confidence factors. For force fields, the *inherent, learned,* and *position confidence* can all be derived from the field map. For the virtual path method they are calculated separately. To obtain a single *confidence factor* for the robot, we simply multiply these normalized values for the path segment.

$$C_t = C_p * C_i * C_d * C_u * C_f \qquad \text{(Equation 13.1)}$$

where:

C_t = Total confidence (0-1)

C_p = Confidence from recent pain

C_i = Inherent path confidence (0-1)

C_d = Learned danger confidence (0-1)

C_u = Position uncertainty confidence (0-1)

C_f = Fitness confidence (0-1)

For the force field method, we have only to multiply the value derived from the worst field strength under the robot's uncertainty footprint by the recent *pain confidence* and the *fitness confidence factor*. It is important to realize that at any one time, most or all of these factors will be one.

The *total confidence* of the robot is then multiplied by the programmed velocities, accelerations, and torque limits to derive operating values. This value is also a useful number to display as it provides an overall assessment of the robot's operation.

Summary

The importance of these concepts is difficult to overestimate. By making these simple calculations a robot can be made to be responsive to its environment, its history, and its condition, and to operate at an optimal combination of speed and safety. Yet again, we have seen the importance of the uncertainty estimate, as well as the power of simple fuzzy logic concepts.

CHAPTER **14**

Becoming Unstuck in Time

In Kurt Vonnegut's classic book, *Slaughterhouse-Five*, the hero is subject to incredible horrors as a prisoner of war in Dresden, Germany during the infamous firestorm raids by Allied bombers. To escape he becomes unstuck in time, and shifts between various periods of reality and a fantasy life on a distant planet. Once unstuck, he is never able to go back to thinking of time in a linear fashion, even when he has been liberated. The experienced robot programmer will find the same shift in thinking taking place, though with any luck it will be a more pleasant experience.

We should learn to think of time as just another axis in our position. Since science fiction has dealt with this concept extensively, it should be a fairly natural jump.

Getting past sequential thinking

In the earlier days of mobile robot development, affordable sensor systems were limited to sonar rangers and simple retroreflective light beam detectors. Given the very limited information these systems provided at any one moment, robots needed to gather information for some interval in order to obtain enough data for analysis.

With the advent of GPS and excellent lidar systems, rich streams of data are available at relatively short time intervals. Today's programmer must recognize that the need to manage time is still critical, and that only the scale of the time frame has changed.

Agent arbitration issues

In order for a mobile robot to operate robustly in real-world applications, the programmer cannot afford to think in sequential time. That is to say, a robot should not simply read data, react to it, and read more data. When the robot is running under some conditions it may appear to be doing just this, but it must be capable of gathering information from different systems and perspectives and then applying the implied corrections to the current moment.

Take for example the hypothetical case of an outdoor robot running with lidar and GPS. Assume that the robot is approaching a tunnel where the GPS will drop out and the lidar must be used for navigation. Up to the transition point, the GPS may have dominated the navigation process especially if there were no suitable lidar features.

There will always be some difference between the GPS and the lidar position indications. During the time when both systems are reporting, it is critically important to find the true position and heading because of the narrower confines of the approaching tunnel. If the GPS is reporting 10 times per second, and the lidar is reporting only 3 times per second, then the GPS may dominate navigation until it drops out completely, without giving the lidar much of a chance. This would not be good, because if the lidar can see the inside of the tunnel it will have a much higher accuracy in lateral position and azimuth, and proper orientation in these degrees of freedom is critical for safely entering the tunnel.

To understand why this is true, think of how GPS implies heading corrections. In the simplest method, two GPS readings are used to calculate an implied path as shown in Figure 14.1. The error between this path and that implied by odometry is the implied heading correction. In Figure 14.1, the heading implied by the GPS is compared to the true path, showing us just how misleading this technique can be.

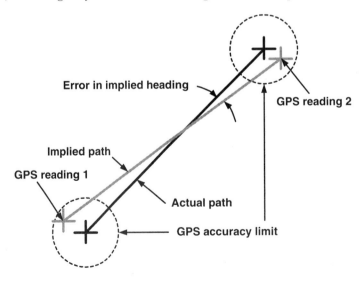

Figure 14.1. Implied heading from two GPS readings

Even given reasonably good dead reckoning and an appropriate fuzzy *agent* for the GPS, there is going to be some amount of disagreement between the two systems. Therefore, it may not be advisable to immediately accept every implied correction from the GPS, because every time we accept a correction to the odometry we may invalidate the data being compiled by the lidar *agent*[1].

Instead, we may wish to give the lidar time to report and be processed before accepting any lateral or longitudinal correction from the GPS. By the time the lidar reports (or runs out of time), the GPS may have two or even three implied corrections. We don't want to lose the richness of this data, but it is not necessary to use every reasonable fix as it comes in.

Here we add a concept to the fuzzy navigation approach discussed in Chapter 12. This factor is the inherent quality of each sensor for each degree of freedom. This *intrinsic quality* is then multiplied by the quality of the fix for its respective degree of freedom. By giving the lidar a higher intrinsic quality, we are telling the arbitration logic that we trust the lidar more for heading and lateral position than we do the GPS.

Thus, our navigator task might be made to collect several implied corrections from the GPS and at least one from the lidar before making a decision on which one to believe. If we *snapshot* each of these implied corrections, and then compare their qualities, we can give slower but more accurate sensors a chance to compete with faster and less accurate sensors.

Data snapshots must include the full position and uncertainty estimates at the moment of the data collection, along with any other information required. As long as no other corrections have been made to the axis during the collection and processing period, we can assume that we are still out of position by the same relative amount we were when the data was collected. Therefore, when the *relative correction* to an axis is calculated, this correction is applied to the present position estimate.

Issues of navigation data latency

There is also the issue of data latency. For example, if our GPS system uses a serial interface to communicate its results to the robot, there will be a certain latency between when the reading is made and when the navigation agent processes it. If

---

[1] It is possible for an agent to translate incomplete data by the amount of the correction made by another agent, but in practical terms I have found this to add noise to the data, especially in cases where it occurs several times during collection.

the agent does not compensate for the movement of the robot during this period, then the correction will be in error by that amount.

In cases where the sensor collects an array of readings, each reading must be compensated for the movement made since the collection started. Figure 14.2 shows the angular distortion that can result from failing to compensate for motion that occurred within the sweep of a lidar.

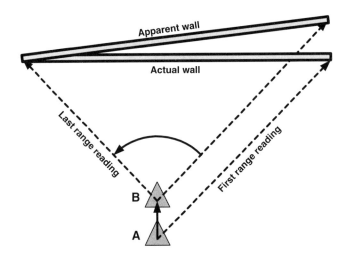

Figure 14.2. Angular distortion from data latency

In this case, the lidar is scanning at 0.3-second intervals (typical for high-resolution lidar), and the robot is moving at about 20 km/hr (13 miles per hour). Only 90 degrees of the 180-degree scan fall upon the wall to be mapped, meaning that the time difference between the first reading and the last reading is .15 seconds. At this rate the robot travels between points A and B or about .83 meters (3 feet) during the time the beam is on the wall. If the range readings are not compensated individually for movement, the measured azimuth of the wall will be in error by 6.7 degrees! At an order of magnitude slower speed, the distortion would still be unacceptable. Worse yet, our fuzzy navigation techniques are unlikely to filter out these errors because they represent a constant bias at any given speed and not random noise—and this is only the angular error!

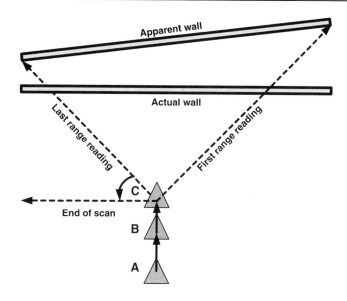

Figure 14.3. Longitudinal distortion caused by uncompensated motion

Normally the data from a scan will be collected and processed by a lidar system and then transmitted as a burst. This means that in addition to the time required for the beam to scan across the wall, three additional time periods will occur from the time that the beam finishes scanning the wall and the time the navigation agent receives it. These times include the .075 seconds required for the scan to reach its end (180 degrees), the time for the lidar to process the readings, and the time for the message with these readings to reach the requesting agent. During this time, the robot will have traveled the additional distance from B to C.

Compensating sensor data for motion

Obviously, we cannot afford to ignore the error caused by motion. There are several levels of compensation that can be made. To some extent, this will depend on whether the lidar processing is done in the same computer as the odometry or in a separate computer. If computing for sensors is done in the same computer as the odometry, the system is said to be *tightly coupled*. If the processing is done in a separate computer, then the processing is said to be *loosely coupled*.

Certainly, the advantage to tight coupling is obvious in that the current odometry estimate is available to the navigation agent in real time; the disadvantage is one of flexibility. If all navigation agents and sensor interfaces must be built into the main computer of the robot, then it may be difficult to expand the system as time goes on.

I have read papers that dismissed loosely coupled systems, explaining that loosely coupled processes cannot adequately compensate for motion distortion of sensor data. We did not find this to be true, at least for indoor robots.

Cybermotion robots were designed for maximum flexibility, so they were loosely coupled. In order for the slave processors to know the robot's current position estimate, a serial broadcast message is transmitted ten times per second. This message contains not only the coordinates, but also the acceleration and velocity of the drive and steering servos. The received broadcast is placed in local memory along with a time reference from a local time counter.

When the lidar data is received, each range reading is converted into a relative XY target position by first calculating the position of the robot at the time of the reading and then adding the vector angle and range to that position. This compensation was completely adequate for all conditions except rapid turning and deceleration. During these types of maneuvers the errors are not constant as they are in straight-line motion, so the fuzzy logic filters them out very nicely.

It may be tempting to dismiss message latency if one is using a fast Ethernet connection between the processes. It is important not to confuse the bandwidth of the connection with the message latency. For small amounts of data, an Ethernet connection may not be faster than a high-speed serial connection.

For high-speed outdoor vehicles, there is no doubt that *tight coupling* is nearly essential. It may become necessary at these higher speeds to record the actual position estimates of the robot into an array at the same rate that the lidar is collecting range readings. Then when the scan data is returned, the backward processing of the array against the laser range data can be accomplished.

Thinking of a mobile robot as multiple robot time-places

We tend to think of problem solving from a static perspective. This implies that we expect the robot to be able to determine its position and heading without moving. This is often not the case. Instead, we must take advantage of our odometry accuracy to allow us to take readings from two or more time-places and assume they are in the same coordinate system. Again, this is only possible with good odometry.

As we have seen, even a GPS system does not offer us an azimuth estimate without moving. True, if our vehicle was fairly long we could mount receivers on the front and back to obtain a heading estimate, but what we are really doing is establishing two simultaneous *time-places*.

For instance, if a robot is traveling down a hall using lidar or sonar it may receive frequent lateral and azimuth corrections, but no longitudinal corrections for some distance. Eventually it should reach a point where it can image a feature that indicates its longitudinal position. Therefore, odometry is the key to gluing together these cues about the environment. In this case, it is necessary that the odometry be accurate enough longitudinally (distance traveled) to keep the position estimate intact.

If your goal is to develop a very inexpensive, simple, or small robot, then you will probably find lidar a bit pricey or bulky. The more limited the sensors available, the more aggressively this concept of time-places must be embraced.

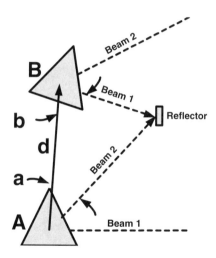

Figure 14.4. Determining the position of a retroreflector from simple beams

Flashback...

This story is a prequel to the story in Chapter 6. A few years before the first practical lidar systems were introduced, we were working with a Navy group to find ways to navigate rack warehouses. It quickly became apparent that the posts of the racks were among the only dependable features from which to navigate. The problem was that there was too much clutter in the area of these posts to depend on imaging them by themselves. We then decided that we would decorate the posts by mounting reflective tape on them, and navigate from this easily discriminated artificial feature (called a *fiducial*).

The problem at the time was that the only affordable sensors were optical readers that would close a circuit when their beams detected a retroreflector. Since the readers returned no range information, it was necessary to imply position information from multiple robot time-places.

We mounted two beam modules on each side of the vehicle: one pointing directly to the side, and one pointing out about 45 degrees diagonally from the robot's forward-direction, as shown in Figure 14.4. When the diagonal beam detected a reflector (time-place "A"), the agent would record the odometry values for x, y, and azimuth. When the lateral beam subsequently detected the reflector, time-place "B" was also recorded. At that point we knew (in terms of our odometry) the vector distance d between points "A" and "B", and we knew the angles at which the beam was projecting at each detection.

The internal angles a and b could then be calculated from the azimuth of the vector between "A" and "B," and the projection azimuths. Given the single-side length d and the two included angles, we could solve for the position of the reflector in terms of the odometry frame of reference.

Knowing a single reference point is not, however, adequate to provide a position or heading correction. Therefore, the robot would save the implied position it had calculated for the reflector and drive on. When a second reflector was encountered, on either side of the vehicle, the robot would repeat the process. Given the apparent position of two reflectors for which the program had provided true positions, the robot's x, y position and azimuth could be corrected. The implied correction from this agent was subjected to all of the fuzzy processing already discussed.

The process then used four robot time-places to calculate a fix, but the fix did provide a correction to all three degrees of freedom. The process worked reasonably well, and the Navy filed a patent on it. As mentioned in Chapter 6, the technique never saw useful service because it was decided to install the subsequent system in a pallet warehouse which had very few places for reflectors to be mounted.

Some years later I was notified that the patent had been granted, but by then our robots were sporting the very capable LMS lidar systems. The story of robot development is one of continuous improvement in the enabling technologies, and therefore continuous change in the robots that use these technologies.

Managing the time dimension

We have seen that the time dimension can work for or against us. We must constantly think of the effects of other things that may be occurring concurrently. We must also ensure that data has been appropriately compensated for latency and collection times.

The consideration of latency is not limited to navigation; for example, a security robot will tend to use its camera in many ways in order to maximize its coverage. When the robot is performing routine patrols, the camera will often be given coordinates to *watch*. Similarly, if an intruder is sensed it will be directed to track the suspect. There are several latency issues here. For example, the sensor data is always a bit old due to processing, the communications system will require time to relay the control message to the camera, and the camera will require a finite time to move to the requested position. To compensate for these delays, the robot must dynamically calculate a "lead" from the current position command according to the target's apparent motion. If this is not done, the camera will be largely ineffective.

If we try to ignore tiny time delays, and the things that can happen during these delays, we may pay dearly in the performance of the robot. On the other hand, if we imagine that we have data from multiple robot time-places available, we can imply a rich amount of information from relatively sparse data. The trick is to begin to think in terms of time and to learn to avoid static and sequential problem solving. We must become unstuck in time.

Programming Robots to Be Useful

In Chapter 7 we discussed the debate between those who believe robots need to be self-teaching, and those who believe in preprogramming. From a business standpoint, a robot that can go to work without installation is very appealing since it can be churned out in ever-increasing numbers and shipped to the far corners of the market without the need for an installation infrastructure. In business speak, it is said to be "scalable."

As we have discussed, this model assumes that the application requires no coordination with door openers, elevators, alarm systems, or other equipment, and can be handled in much the same way in every installation. It also assumes that the task is relatively simple and repetitive.

In the end, there will be applications in which self-teaching is possible, and applications in which it is not. The concepts presented in this book are directed toward robots that can perform complex and varied tasks, so we will take it as a given that some form of instruction must be provided to the robot.

Preprogramming vs. teaching paths

For applications that must undeniably be programmed, the almost universal consensus is that the easiest manner of programming would be to walk behind the robot teaching it the route and its duties. I said "*Almost* universal consensus" because the few of us who disagree with this approach are those who have actually tried to use it!

Using a 500-pound mouse

Cybermotion's first programming method was walk-behind teaching. Granted, the technology was still rather primitive at that time, but the lessons were fairly obvious.

The first problem was one of position accuracy. We quickly learned that recording "bread crumb" paths was a bad idea for indoor robots, because any unintentional weaving would be recorded and duplicated by the robot. The size of the data required for bread crumb paths was also orders of magnitude larger than for other methods.

True, the data generated by walk-behind teaching could be massaged by line-fitting programs to generate smoother curves and straighter lines, but there is still a problem. Anyone who has used a drawing program that performs curve fitting (as lines are drawn using a mouse) can quickly identify with the problem of this approach. The algorithm often makes wrong assumptions about the artist's intentions.

For example, under most circumstances a robot running along a wall will maintain a constant distance from the wall, so it is logical for the learning process to attempt to massage the bread crumb data to accomplish this. However, there are times when the path must violate this assumption. To communicate these subtleties to the learning program, the walk-behind teacher is essentially trying to accomplish the same thing as the aforementioned artist, but with a 500-pound mouse!

Distance issues

Another issue that is often not appreciated is the sheer size of some environments. Many of the installations we were to program were more than a million square feet in size with tens of miles of paths. In most applications, there is no return on investment by putting a robot into a small area, so all but the most modest installations represented a marathon effort to program.

The choice is between walking twenty miles while arguing with a software interface about what you really wanted to teach it, or sitting at home and drawing paths on a map. It should be obvious that the walk-behind approach is not much fun. Road-following vehicles that can be ridden while being taught are another matter however.

Data management issues

In Chapter 7, we discussed the concept of building maps from scratch vs. starting with a high quality drawing. A good facility drawing has information about much more than just walls and doors, and shows things that no mapping program can determine. Any good automation engineer knows the golden rule:

Never throw away useful information.

In Cybermotion's PathCAD programming environment, one of the many objects that can be placed on the map is the location of radio equipment and interface hardware. Dropping a radio symbol onto a map automatically shows the estimated coverage areas. Moreover, there is a maintenance advantage to such data. Since radios and interface equipment are often placed in equipment closets and in overhead ceiling spaces, it can become an expensive Easter-egg hunt to try to find them for maintenance. This kind of utility is only possible when using proper building drawings.

As another example of data-management issues, destinations must be named in logical ways for future reference. For a walk-behind programmer to logically assign these names would require significant graphical interface with map-viewing software.

Coordination issues

Finally, there are issues of coordination. If robots become smart and dexterous enough to open doors, punch in access codes, and operate elevators, then some of the issues of coordination may be relieved, but others will always remain.

Today, most industrial customers will not accept stand-alone systems of any sort. Some of the first questions asked during the sales cycle will usually involve integration methods. For example, a factory material-handling robot must be capable of interfacing with pick up and drop-off points. If it depends on people at each end, it will suffer significantly in efficiency. A security robot must be capable of responding to fixed alarm systems, and disabling these alarms to avoid false alarms when it moves into their protected zones.

Specifying these interfaces during walk-behind teaching is very problematic. For all of these reasons, the only significant justification for this type of teaching involves road following vehicles where the programmer can drive the vehicle during the teaching process. The focus here will be on off-line programming methods.

Embedding data into maps

Both of the remaining methods of robot programming involve embedding data into maps. This can be done in the professional versions of programs such as AutoCAD and IntelliCAD[1]. The two most common embedding languages are *LISP* and *VBA*[2] (*Visual Basic for Applications*). In LISP, graphical objects are called entities, and in place of properties these entities have *data lists*. Thus what may appear as a simple graphical object, such as a circle, may have much deeper meaning as defined by its *type*.

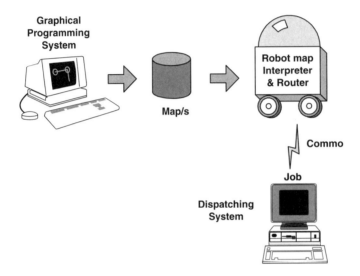

Figure 15.1. Program data flow for a map interpreter system

---

[1] AutoCAD is a trademark of Autodesk, Inc.; IntelliCAD is a registered trademark of the IntelliCAD Technology Consortium.

[2] VBA is a trademark of Microsoft Corporation.

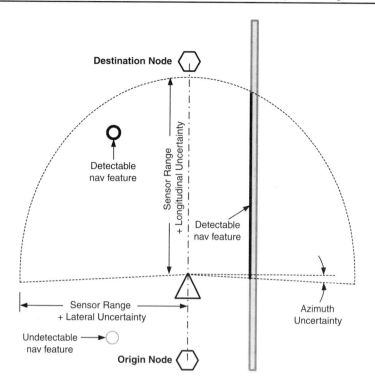

Destination Node

Detectable
nav feature

Sensor Range
+ Longitudinal Uncertainty

Detectable
nav feature

Sensor Range
+ Lateral Uncertainty

Azimuth
Uncertainty

Undetectable
nav feature

Origin Node

Figure 15.2. Map interpreter searching for navigation features

Map interpreters

One method of conveying instructions to a robot is to provide it with a map of the environment into which symbols have been embedded that indicate destination nodes and navigation features. As discussed in Chapter 7, *nodes* are the waypoints by which paths are defined. The robot therefore receives commands (jobs) to drive to various nodes.

In this programming scheme, the robot interprets the map by graphically searching outward from the robot's position estimate, looking for nearby navigation symbols that its sensors might be able to detect. The appropriate agent for each feature type is activated and provided with the properties of each feature it may be able to detect. If these features are located, the navigation agent calculates the implied correction as discussed in the preceding chapters. In this way, the robot navigates from node to node, effectively using its position estimate as a graphical version of a program pointer.

Figure 15.2 graphically depicts the area over which a robot will try to find navigation features. Notice that the arc is not the lidar sensor's +/– 90-degree angular range, but rather 90 degrees plus the azimuth uncertainty and –90 degrees minus the azimuth uncertainty. Also, notice that the arc is elliptical and not circular due to the lateral and longitudinal position uncertainty[3].

Events and targets

Actions that the robot may need to perform along a path can be indicated by dropping *event objects* onto the map along the path. One of the most common actions a security robot might be programmed to perform is to "watch" a place with its camera as it passes within view. In the example shown in Figure 15.3, two events have been dropped on the path segment, each with its own properties indicating the actions to be taken.

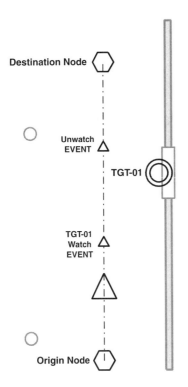

Figure 15.3. Programming events with a map interpreter

---

[3] If the sensor range is very much greater than the maximum expected uncertainty, then it may be reasonable to ignore the position uncertainty in setting the search window.

As the robot passes the first small *event* triangle in Figure 15.3, its map interpreter will assign the camera to watch the coordinates of TGT-01. A "target" is simply a defined set of coordinates. When the robot passes the second event, the camera will be freed to return forward or perform other operations.

Notice that in this example the event triangle tells the robot to execute this event only if its direction agrees with that of the event triangle. Since the robot might be traveling either direction on the path, it is necessary for events to be directional in nature.

At first, it might seem that the two events above could be made bidirectional in nature—that is, the first event encountered would always start the watch and the second would end it, no matter in which direction the robot was traveling. In actual practice, however, events and actions are rarely symmetrical. For example, to allow for the time it takes the camera to swing around onto the *target*, the event to start watching a coordinate usually occurs further from the target than the event that terminates the watching behavior.

Events are one-shot occurrences—that is, if the robot has passed an event and suddenly gets a longitudinal correction that indicates it has not yet reached the event, the event is not executed the second time it is "reached." This is also the case if the robot is forced to back up in a circumnavigation maneuver and subsequently re-crosses the event. Also, since it is very unlikely that the robot will move over the exact position of an event, the firing of an event will occur when the robot is abreast of the event position.

Actions and action nodes

To understand the concept of *action nodes,* it is necessary to move beyond thinking of the robot as simply going from place to place, and to think of it as going from state to state as well. For this reason, it is helpful to think of the act of moving from one node to another as simply one type of *action*. In this way, we free ourselves from the rigidity of thinking only in terms of movement.

For example, there is a significant change of state that takes place while a robot is at its charger, even though the robot does not move as this occurs. If we define nodes in terms of place-states, then the same router logic that sends a robot from one place to another can easily direct it to change its state.

Our robot in the previous examples may be moving toward its destination node as one of several waypoints required to get to a final destination, or it may be intending to stop there to perform a task. For example, a material-handling robot may be going to the destination in order to accomplish a pick-up or drop-off of materials. Nodes can be specified as different types according to the task to be performed. A given place may be used as a waypoint some of the time, and as a place to perform tasks at other times.

The concept of an *action node*—sometimes called a *virtual* node—is that it resides at the same position as an ordinary node, but that to "get" to it the robot must go through certain actions. These actions can include picking up a load, charging its battery, or performing security tasks. Similarly, to get from the action node back to the ordinary node the robot may need to take other actions. At this level, the map interpreter approach begins to become cumbersome for all but the most limited number of actions.

Advantages and disadvantages of map interpreters

The biggest advantage of the map interpreter approach is that the map contains virtually all the program information. The data structure is therefore very clean. The incorporation of force field concepts into the map approach is also very straightforward.

One potential disadvantage of the map interpreter approach is that it requires much more onboard processing power and complexity than the programmed approach that we will discuss next. Since so much heavy processing is being performed onboard, it can also be more difficult to understand exactly what the robot is thinking at any given moment, making remote diagnostics somewhat more challenging.

Another disadvantage to the map interpreter approach is its inflexibility. Unless a scripting language is patched onto the map interpreter, the robot can only perform tasks that were anticipated when its map interpreter was created. If a new type of task is required, or if an old task must be performed in a slightly different way, then the map interpreter must be modified to include these event or action node objects. If a scripting language is patched onto the interpreter, the technique loses its advantage of data conciseness.

Maintaining a fleet of robots becomes increasingly difficult as their core software begins to fragment into application specific versions.

Text-based programming

Perhaps the most conventional approach to programming robots is through the use of text-based programs. Before being sent to the robot, these programs will normally be compiled from text into a *P-code* (Pseudocode) format that is more concise and easier to interpret[4] than text. Note that conversion of text programs directly to the native code of the robot's computer is generally an order of magnitude more difficult to implement and of little advantage.

The first step toward programming robots in this way is thus to develop a robot language. This step is necessary, whether the programming is to be done with a keyboard or graphically, as it is the means by which the robot is instructed. Graphical programming can be made to generate P-code directly, but then a disassembler will be required if the resulting programs are to be viewed. We will assume that the programming language will start with text-based programs. In this way, the results of any graphical or automatic programming process can be viewed with an ordinary text editor, and subsequently modified if necessary.

In conventional programming languages, most instructions are executed sequentially. Each instruction is completed before the next is begun. A robot control language may look very much like a single threaded type of language, but it is inherently much more parallel (multitasking) in nature.

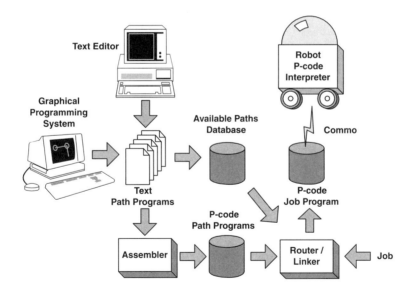

Figure 15.4. Simplified programming data flow for P-code language

---

[4]   See Chapter 2 for a discussion of Pseudocode compilers.

When a conventional program enables an interrupt, the main thread can go on to do other work while an interrupt task collects the requested data from a serial port. A robot control language will contain many instances similar to this. The difference between a conventional language and a robot control language is largely a matter of degree; a robot control language is largely parallel while a conventional language is largely sequential.

The best way to think of a robot language is to view it as a script that controls behaviors. These behaviors, once started, will continue in parallel with other behaviors and actions directed by the program. It is also crucial that the language be very simple and easy to understand so that the robot can be installed by mortals.

Path programs

As discussed in previous chapters, path (or action) programs are designed to allow the robot to move from one *destination node* or *action node* to another. Each destination node must be listed as the *starting node* for at least one path program, and as the *ending node* for at least one path program. If this is not the case, the robot will be unable to either go to a node or to leave a node once it has arrived there.

The paths represented by these path programs are typically quite short, and in some cases begin and end at the same physical position. The key properties of a path program are its starting node, its destination node, and its costs[5]. A path program may be generated by a graphical interface or a text editor as shown in Figure 15.4. Once generated, a path program can be edited as required. The program is then compiled into a P-code and its properties entered into a database for use by a router.

While a *destination node* implies a place the robot can go, an *action node* implies some task to perform at a destination. A robot can then be given a *job* by simply specifying one of these kinds of nodes.

It is the function of the router to put together the lowest-cost sequence of path programs (in P-code) to take the robot from its current node to the desired destination. Once this program is compiled it is called a *job program*. Job programs are sent to the robot, which can execute them with no further interference on the part of the router as long as the path is navigable.

---

[5] See Chapter 14 for a discussion of routing and the costs associated with paths.

As an example, the following instruction would cause a Cybermotion robot to drive by dead reckoning at a speed of 2.50 ft/sec from its current position to the coordinates *x* = 10.00, *y* = 20.00.

```
RUN 250,1000,2000     ;Run to 10.00/20.00 at 2.50 ft/sec
```

If the robot is to use a wall as a navigation reference, then the wall must be specified before the robot begins to run. In the following example, the robot is to use a wall running parallel to the path, five feet to the left. Note that units of distance are in 1/100ths of a foot, and the value –500 implies 5.00 feet to the left.

```
WALL    –500          ;Look for wall 5.00 ft. to left
RUN  250,1000,2000     ;Run to 10.00/20.00 at 2.50 ft/sec
```

This example actually activates two concurrent behaviors. The RUN behavior is a driver that attempts to take the robot from the current estimated position to the specified position. The WALL behavior activates a navigation agent that will correct the robot's lateral position and heading estimates as it drives.

Note that the WALL function assumes that the robot is going to run parallel to a wall. With sonar-guided robots, this is nearly always the case since sonar range is so limited. For lidar navigation, it may be desirable to permit navigation at an oblique angle to a wall. In this case, two points on the wall, or a base point and an azimuth must be specified.

Unfortunately, the use of numeric values in programs causes two problems: first, it is not obvious what the value represents, and second, it can be very difficult to change a value across a number of path programs.

In the previous example, the coordinate *x* = 10.00, *y* = 20.00 is a node. If we decide to change the position of this node, we will need to search all path programs for these *x/y* values. For this reason, it is much better to use symbols, which can alleviate these problems.

```
Defp  Node_AB,1000,2000   ;Define place Node_AB as 10.00/20.00
Defc  Fast, 250           ;Define constant fast as 2.50 ft/sec
Defc  LeftFar, 500        ;Define constant LeftFar as –5.00 ft.

WALL    LeftFar           ;Look for wall at LeftFar distance
RUN  Fast, Node_AB        ;Run to Node_AB
```

In the above program, we have defined the symbols Node_AB, Fast, and LeftFar. But these symbols should be global—otherwise they may be defined differently in other path programs. To do this, we need to generate definition files, and then *include* them into path programs. Once this is done, we have only to modify a position or constant definition in one place and recompile all path programs that might use it.

Our path program might begin to look as follows:

```
Include Local.def      ;Constants such as speeds and distances
Include Map.def        ;Place constants such as Node_AB

WALL   LeftFar         ;Look for wall at LeftFar distance
RUN  Fast, Node_AB     ;Run to Node_AB
```

The above language is rudimentary but very flexible. The Cybermotion language contains over 100 different instructions like WALL and RUN. Among these are instructions for reading and writing data to various blackboard memory locations. In this way, the language maintains the ability to modify virtually any behavior of the robot in the field.

Conditional instructions

Conditional instructions can also be included in this type of language. For example, the *event* instruction in the Cybermotion language specifies a set of coordinates on the path; it also specifies the number of instructions that follow the event instruction. These *conditional instructions* are to be executed when the event position is reached.

```
WALL LeftFar                   ;Look for wall at LeftFar distance

DO@ 1,EV_0                     ;Execute the following 1 instruction at EV_0
  WATCHXY Indef, TG_01         ;Watch TG_01 indefinitely (until cancelled)

RUN Fast, Node_AB              ;Run Fast to Node_AB
```

In the above example the robot will navigate to Node_AB using a wall on its left. As it crosses the event position EV_0, it will begin watching the coordinates defined as TG_01. Therefore, our program initiates three initial behaviors: one for driving, one for navigating, and one that waits for the robot to reach EV_0. Once the robot passes abreast of EV_0, the event "fires," and a new camera behavior is started to cause TG_01 to be watched.

Conditional instructions can include not only events, but also tests of variables. For example, an instruction called **DOWHEN** is used to test a variable and to execute conditional code the first time the test becomes true. Likewise, an instruction called **DOWHILE** tests a variable and repeatedly executes a list of instructions as long as the test is true.

Advantages and disadvantages of text programming

The type of programming language just discussed has the advantage of being easily viewed, tested, and modified, and of being extremely concise. The actual P-code *job programs* are usually between 10 and 500 bytes in size, making them very quick to transmit. Most importantly, the language is extremely flexible.

The biggest disadvantage to programming with a text editor is that it is labor intensive and there is a tendency to make typographical errors; for example, there may be hundreds or even thousands of position definitions. If each of these must be read from a map and typed into a definition file, the process will be time consuming and error prone. If this type of drudgery can be eliminated while maintaining the inherent transparency and flexibility of text programming, then we can have the best of both worlds.

Graphical generation of text programs

For systems that require the flexibility of a scripting language, the best approach is therefore to generate programs using a graphical programming tool. Cybermotion developed the PathCAD system to perform this task, and I will use it as the example here. This system can instantly and flawlessly maintain definition files, write path programs, and even check for common programming deficiencies and mistakes.

The graphical programming process allows the same degree of flexibility as hand coding, but at a much lower labor cost. We found that the process of writing and debugging programs graphically was between 10 and 30 times faster than hand coding. Much of the time saved was in the debugging process, since most common mistakes were eliminated during programming.

Creating features with embedded properties

The process of generating text programs graphically starts by drawing path lines and by embedding navigation objects and cues into the map drawing. These objects are created using the *navigation object* creator, and include nodes, range lines, target positions, event positions, and so forth.

Once they are embedded, these features are used in several ways. As shown in Figure 15.5, a definition file creator program can automatically extract the names and properties of features to generate the definition files discussed earlier. For example, the position of a node on the drawing can be used to generate a "define place" statement for the node.

Creating path programs

Path programs are next generated in a semi-automatic manner by picking a starting and ending node. The path program creator then automatically opens the program, and adds a header telling where the path goes from and to, who programmed it, and when it was created. The appropriate *include* statements are appended next. The body of the path program is then built by selecting the various functions to be executed along the path in program order. This is done by selecting functions from menus of those available. Each function will represent a step in the program.

When a function is picked, it will query the programmer for its needed parameters. These parameters are graphically picked from among the objects in the map. The text program uses only the name of the feature or parameter. The actual value of the feature will be read by the P-code assembler from the definition file whose generation we just discussed.

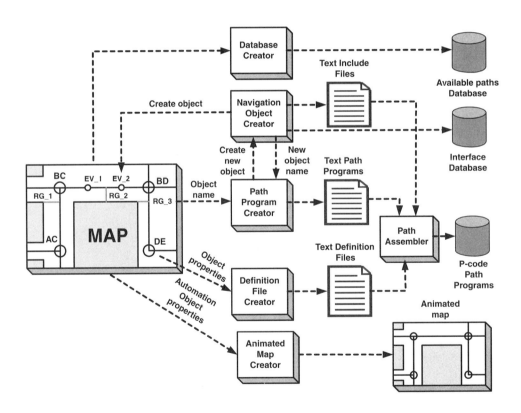

Figure 15.5. PathCAD graphical programming of P-code programs

For example, take the case of starting a watching behavior at an event position. The target to be watched and the event position on the path may or may not have been previously placed on the map. In either case, the programmer selects the "watch event" function from a menu. (Note that functions that are not legal in the current program sequence are disabled, thus minimizing programming errors.)

When the "watch event" function has been specified, the graphical interface will ask for the event position. If the object exists, the operator simply clicks on the object and its name is imported into the program. If the event has not been previously created, the operator enters "n" for "new." The *path program creator* will call the *navigation object creator*, which will in turn create an *event* name and ask where the new event should be placed. Once the new object has been created, the *navigation object creator* will return its name to the *path program creator*, and the process continues.

```
;////////////////////////////////////////////////////////////////////////
;   B13_CDN3.sgv- Path file generated by PathCAD Version 3.63.
;
;   Path program from B13_CD to B13_N3.
;   Map Name: B13 [B13]
;   Map Desc: B13 Warehouse
;
;   Generated by: jmh
;   Generated on: 04/12/03 at: 00:35
;////////////////////////////////////////////////////////////////////////
;   — INCLUDE FILES —
;
    INCLUDE local.def      ;Local operating parameter definitions.
    INCLUDE global.def     ;System constants and addresses.
    INCLUDE B13.def        ;Map definitions from GENDEF.
;
;------------------------------------------------------------------
;
    AVOID   NC, Far,Wide   ;Set sonar, front=Far, side=Wide.
    FID 0,  TG_28          ;Use FID at TG_28 with default diameter.
    FID 0,  TG_26          ;Use FID at TG_26 with default diameter.
    LOOK    FarFwd, FarSide, FarSide ;Look favoring both sides.
    FID 0,  TG_49          ;Use FID at TG_49 with default diameter.
    FID 0,  TG_210         ;Use FID at TG_210 with default diameter.

    DO@ 1,  EV_10          ;Execute the following at EV_10.
     WATCHXY Indef,TG_10   ;Watch TG_10 until canceled by WATCH or PAN.
                           ;DO Group closed.

    RUNON   Fast,   B13_N3
;
```

233

```
;-----------------------------------------------------------------
;    — End of file c:/B13/B13_CDN3.sgv —
;    — Total Path Length = 20.114 feet. —
;    — Worst Dead Reckoning = 0.000 feet. —
;    — NavNanny warning distance = 15.000 feet. —
;
;    NavNanny rates this program 100.000% above adequate (awesome).
;-----------------------------------------------------------------
```

Figure 15.6. Typical program generated by PathCAD graphical programmer

Figure 15.6 contains a typical path program generated by PathCAD. A navigation agent is activated and told to use any combination of four FIDs. FID is shorthand for *Fiducial* or a small reflective target. A camera behavior called "looking" is also initialized to point the camera at anything that comes within the specified range on either side. Notice that comments are automatically generated to aid the viewer in understanding the purpose of each line of code. The presence of these comments makes editing small changes into the programs a simple and relatively safe process.

Generating path databases

At runtime, the router will need to know what path programs are available, where they begin and end, and their cost. The database creator generates files with this information from the path programs that have been completed. The cost of each path is initialized, but may be modified during operation as discussed in Chapter 13.

Generating include files

In some cases, a feature may require control code that can also be generated automatically. This is analogous to a *peripheral driver* in a PC. For example, if an automatic door is dropped onto the map, PathCAD will ask a few questions about how it is interfaced to the system, and then generate special *include* files that can be dropped into the robot's path programs to open and close the door at the appropriate times.

Consider a more complex task. If an *elevator node* is placed on a map using PathCAD, then the graphical programmer will want to access the definition file for the elevator. This file tells the program what kind of elevator it is, whether it has one or two doors on the car, how many levels it serves, and so forth. The first time the elevator is referenced and defined, PathCAD will generate the definition file, include files for calling the elevator, include files for releasing the elevator, and "vertical path" programs running between all its levels.

Generating automated maps

PathCAD generates a map for remote display by the operation software. At runtime, the robot will be shown on this map according to the position and heading estimate it reports. Other objects such as navigation diagnostics, intruder symbols, and other information will also be displayed on the map at runtime.

When automation objects such as doors and elevators are dropped onto the map, PathCAD creates files that automatically open interfaces to the controls of the systems that the objects represent. The status reported by these objects is then displayed graphically on the map. Doors on the map can then appear to swing or slide open, depending on the status read from the interface.

Figure 15.7 shows an animated door display; the door is named "dock door," and it allows the robot to enter a shipping area where its charger is located. Notice the small circle on the middle of the door. This circle represents the control point for the door. If the operator clicks on the control circle, a small control panel for the door is displayed as shown in Figure 15.8.

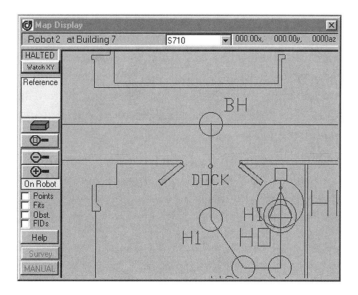

Figure 15.7. Runtime animated map display of door
(Courtesy of Cybermotion, Inc.)

235

Figure 15.8. Door control panel

Notice that the control panel inherits the name of the door (dock), and all of the necessary interface information required to control and monitor the door. All of this functionality was provided by the graphical programming system when the door symbol was dropped onto the map. The runtime control software (i-Con) automatically inherited the necessary information to provide the functionality.

In fact, every animated object that appears on the runtime map can be clicked by the operator to obtain its status and/or to control it. In the case of the SR-3 security robot, these objects include the robots, elevators, alarms, tagged assets, intruders, flames, and so forth.

Expert assistance

I have never been fond of the term AI (artificial intelligence). My personal disdain for the title is philosophical, and has its roots in the fact that it seems to imply awareness or sentience. Society has generally concluded that we as humans are sentient, and that we all tend to experience what we call "reality" in much the same way, while the creatures we choose to ruthlessly exploit are deemed to be fair game because they are not capable of such perception.

In fact, we do not really know anything except what we personally perceive to be reality. We can never truly know if others share this perception. Because of this, and Turing[6] tests aside, I cannot imagine how we would ever know if we had created "artificial intelligence."

---

[6] Alan Turing's 1950 paper, *Computing Machinery and Intelligence*, published in the journal, *Mind*, held that computers would eventually be capable of truly thinking. Turing went on to propose a test whose basis was that if a blind observer could not tell the typed responses of a computer from those of a fellow human, then the machine was "thinking." The fallacy of this proposal is in presupposing that the human respondents would be thinking. The unlikelihood of this is easily demonstrated by visiting an online chat room or listening to a radio talk show.

Having ranted against AI, I will say that I personally like the term *expert system*. This term simply implies that the expertise of someone who is skilled at a task can be made available through a software interface. Since robotic systems tend to be complex in nature, expert systems can be extremely useful in both their programming and operation.

PathCAD has several expert systems built into it. These experts watch the programming sequence for situations where their expertise can help in the process. The most vocal of these is called "NavNanny." NavNanny watches the programs being created to assure that they contain sufficient navigational feature references to assure proper operation.

To accomplish this, NavNanny estimates the narrowness or tightness of a path from the features and settings it contains. It then estimates the distance that the robot could drive without a correction before risking becoming dangerously out of position. This distance is called the *dead reckoning distance*. Notice that in the example of Figure 15.6, PathCAD has attached a footer to the program which records these figures. The path assembler electronically reads this footer and uses it to calculate the path's estimated time and risk cost.

Conclusions

In some jobs robots can be self-teaching; however, in more complex applications they must be preprogrammed. Programs may be implied by objects embedded in a map, or they may be discrete from the map. Discrete programming can be done traditionally or graphically, or with a combination of the two. The application will dictate the most effective programming approach.

CHAPTER 16

Command, Control, and Monitoring

Many robot designers have an innate longing to create truly autonomous robots that figure out the world and "do their own thing." This is fine as long as the robot is not expected to provide a return on investment to anyone. Like most of us, however, most robots will be required to do useful work. For these robots, all but the simplest tasks will require management and coordination.

Once we have developed a control language and a P-code router or a map router, we have the means to specify *jobs* for a robot. A job is specified by its ending node. For a P-code driven robot, this will cause the router to create the job program by concatenating the lowest cost combination of action programs that will take the robot to the destination. If the robot is to perform some task at the destination, then the task action will be specified implicitly in the action program ending at the destination node (as discussed in Chapter 15).

For a map-driven system, the destination node is sent directly to the robot, whose onboard router plans the route to the destination. As discussed in the previous chapters, a map-driven system does not inherently know how to perform actions other than movement. Therefore, in order to support the robot doing useful things once it arrives at its destination, either the action must be implied in the properties of the destination node, or the map interpreter must have a scripting language.

Once our robot has the ability to go somewhere and do a job, the obvious next question is: "Where will jobs come from and how should they be managed?" Like all of the previous choices in our architecture, this choice will depend heavily on the application for which we are designing the robot. The easiest way to place this subject in perspective is to start at the least complex end of the spectrum.

Unmanaged and self-managed systems

The two most common examples of unmanaged systems are area coverage systems and loop systems. The reason for the popularity of these systems has largely been that they stand alone and do not require a communications and control infrastructure.

In the past, installing and maintaining a network of radio transceivers significantly impacted the overall cost of a system and made it more difficult to justify. With the popularity of 802.11 Ethernet radio networks and the proliferation of computer workstations, it is now possible to ride on existing corporate or government customer systems with little or no additional hardware cost. For smaller business customers and high-end consumers, who do not have existing radio systems, the cost of a single 802.11 access point is low enough to be practical in many instances. With the cost barrier eliminated, unmanaged systems are already beginning to disappear for most high-end applications. Even so, it is still useful to discuss how unmanaged systems work, as certain concepts can be useful in managed systems.

Area coverage systems

Area coverage systems are most commonly seen in cleaning applications. As we have already discussed, these systems must be initialized and managed by a human worker on premises. For domestic cleaning this is no problem, but for commercial cleaning the need for an on site operator clouds any cost justification based on labor savings.

In these commercial applications, the operator will be expected to provide menial work in parallel with the robot. Since this implies a minimal wage job, it is apparent that the robot's operation must be very simple. It is equally important that the tasks for the robot and operator take roughly the same amount of time. If this is not the case, then either the robot or the operator will be idle while waiting for the other to finish a task. Trying to assure a potential customer of the savings of such a system may be a challenge.

Loop systems

Loop systems are most commonly seen in mail delivery and very light materials-handling applications. In such applications, the vehicle is loaded with mail or consumable materials and is sent out to execute a service loop. Along this loop the vehicle will make programmed stops to allow the clients to retrieve or load items. In most cases, the system is not sophisticated enough to know if there is mail addressed to anyone at a specific stop, so the robot simply halts at each stop for a predetermined time and then continues on.

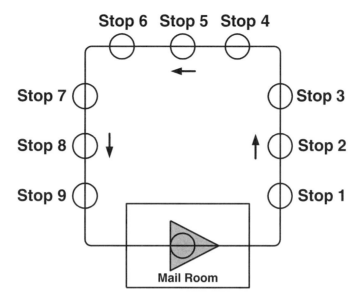

Figure 16.1. Simplified diagram of loop dispatching

The obvious problems with loop systems are their low efficiency, lack of flexibility, and the lack of monitoring. For example, if the robot only has mail for one or two stops it will need to make the entire loop just the same.

One of the methods used to overcome the efficiency problems is for the loop-dispatched vehicle to carry or tow a payload carrier. The loaded carriers are dropped off at predetermined points such as mailrooms or logistics points. The vehicle may then pick up the carrier that it previously delivered, and return to the dispatching center. If the robot arrives to drop off a carrier, and the previous (empty) carrier has not been removed, then either two parking spaces will be needed or the robot must shuffle the empty carrier out of the way before dropping off the full one.

A simple sketch of one such system is shown in Figure 16.2. In this case, the robot is equipped with a lifting device and once it is under the payload carrier it activates this lift so that the carrier is resting on it, and then drives away. While these schemes make loop dispatching more practical, they do so at the cost of system flexibility.

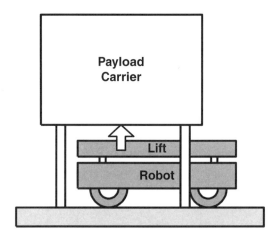

Figure 16.2. Piggyback payload dispatching

While the efficiency problem of loop dispatching is obvious, the lack of flexibility may be an even more significant problem. If a blockage occurs anywhere along the loop, the system is disabled completely. If the vehicle is disabled along the way, there is no inherent way of knowing about the problem until it fails to return to its starting point on schedule.

Ping-pong job management

One of the simplest techniques of job management is the *ping-pong* method. This technique has been used for light materials handling robots and AGVs (automatic guided vehicles) in a number of applications.

Under the ping-pong scheme, the robot is dispatched from an onboard interface. People at various workstations load the vehicle and then send it on its way using this interface. There are two problems with this method; the efficiency of the robot depends upon rapid servicing at each destination, and like loop systems there is no inherent central control from which its location and status can be monitored.

The most obvious problem is that if a robot arrives at a destination and is not serviced, it will be effectively out of service. If the robot is programmed to automatically return after a period without service, it may do so with its undelivered cargo. In this case, the sender must either unload the robot or send it back again.

Additionally, people at the robot's destination have no inherent way of knowing to expect it. If something goes wrong along the way, a significant time may lapse between when it is dispatched and it is missed. In such an event, it will then be necessary to send a rescue party along the robot's route to locate it. Customers seldom show much enthusiasm for such expeditions.

To overcome the monitoring deficiency, some ping-pong dispatched systems have been retrofitted with central monitoring. The problem with this approach is that it incurs most of the costs of central dispatching with only the single benefit of monitoring. In many cases, the ping-pong method violates the simplicity maxim that states: "*A system's design should be as simple as possible and no simpler.*"

Automation does not loan itself well to halfway measures. Once the decision is made to automate a function it is a declaration of revolution. Either the revolution will spread to include the system's control and to integration with adjacent systems, or it will ultimately stagnate and fail. This is because the pressure to increase productivity never ends. If new ways to improve efficiency are not found on an ongoing basis, the customer will begin to forget the initial gains and look for new solutions.

Dispatched job management

Most commercial applications will therefore continue moving toward centrally managed and dispatched systems. In such systems, the software that manages jobs is usually called the *dispatcher*. The *dispatcher* is normally at a fixed control point, but in some cases may be onboard the robot. The jobs that a robot will be requested to perform will normally come to the *dispatcher* from one of four sources:

1. Internal requests

2. External requests

3. Time-activated job lists

4. Operator requests

Internal job requests

Job requests generated within the robot or its immediate control can include requests to charge its batteries, and empty or load job-related materials. The specifics are again application dependent. A nuclear inspection robot may consume inert gas through its detectors, a scrubbing robot may need cleaning fluid, and a vacuuming robot may need to dispose of waste. None of these requirements is necessarily synchronous with other tasks, so efficient operation requires that they be performed on demand.

External job requests

External job requests also depend on the robot's calling in life. For a security robot, these can include signals from panic buttons, fire alarms, motion detectors, door switches, card readers, and even elevators.

External security jobs fall into two broad categories: alarm response and routine investigation. Alarm response jobs usually direct the robot to move to the alarm site, start surveillance behaviors, notify the operator, and wait for the operator's direction. This is often referred to as *situation assessment*. The purpose is usually to confirm that an alarm is valid and to aid in formulating an appropriate response.

Investigation jobs may simply direct the robot to drive by the signal source to document the situation. For example, someone entering an area after hours may generate an investigation response even if the individual has all the proper clearances. These types of responses require no operator assistance and are simply a way of increasing the richness of the robot's collected data.

For a materials-handling robot, external jobs will normally be created automatically by other processes in the facility. These processes will normally involve either a request for pick-up or a request for drop-off. Since the various processes are usually asynchronous, these requests can occur randomly. Because the robot(s) cannot always perform tasks as quickly as they can come in, there will need to be a *job buffer* that holds the requests. Additionally, the processes from which the jobs are generated will usually be mechanically buffered so that the process can continue uninterrupted while waiting for the requested job to be performed.

In factory materials handling, most processes are controlled by programmable controllers or dedicated microcomputers. In these smart systems, where the station must receive service quickly to avoid downtime, the controllers sometimes do not wait for the condition to occur. Such process controllers often calculate that service will be needed and make the requests before the event, thus minimizing the down time while waiting for service. These systems may even post a requested time for the robot to arrive so that the dispatcher can plan the job well ahead of time.

To date, most cleaning robots have been semiautonomous. These systems have been moved from area to area by their operators and then turned loose to perform their jobs on an area basis. Since these systems do not fully free the operator, it may be more difficult to justify their payback. For this reason, high-end cleaning robots will migrate toward more automatic job control.

Time activated job lists

A security guard often has a route that must be patrolled at certain intervals during a shift. Likewise, a cleaning crew will normally clean in a fixed sequence. In both cases, there will be times when this routine must be interrupted or modified to handle special situations.

Robots that perform these tasks will likewise be expected to have lists of things to do at certain times and on certain days. For a security robot these lists are called *patrol lists*. While guard routes are usually constant, security robots may be programmed to randomize the patrol in such a way as to prevent anyone from planning mischief based on the robot's schedule.

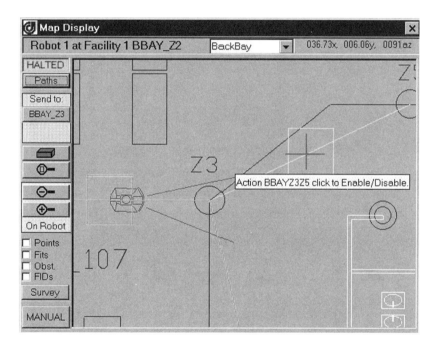

Figure 16.3. Graphical control of the path disabling process

It is not uncommon for job lists to vary by day of the week, or even time of day. For example, the first patrol of the evening by a security robot might need to be sequential so that it can be synchronized with the cleaning staff (or robots) as they move through the facility. Subsequent patrols may be random. Likewise, cleaning robots may clean only one section of the building on one day and another section on the next day.

It is important that these job lists be flexible. For example, if an aisle is blocked for a night, it is important that the operator be able to specify this simply and not be required to create a new job list. In i-Con, we accomplished this by allowing the operator to select a path on the building map and disable it as shown in Figure 16.3. The path then changed color so that it would be obvious that it was disabled. Paths can also be *suspended* for specified periods, in which case they will automatically be brought back into service at a later time.

Operator job requests

Until such time as robots have taken over the entire planet[1], it will be necessary for most of them to do the bidding of humans. In many cases, this means that unscheduled jobs must be performed spontaneously at the request of humans.

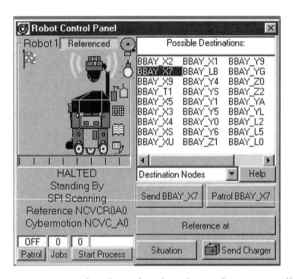

Figure 16.4. Selecting destinations from text lists
(Courtesy of Cybermotion, Inc.)

For a security robot it is common for the console operator to want to put "eyes on" a situation. This may be the result of something viewed on a closed-circuit video, or of a report received by telephone. Similarly, a cleaning robot may be needed when an unscheduled cleaning task arises.

---

[1] The date of such a takeover has been repeatedly delayed. The time frame is still estimated by most robotics entrepreneurs to be about three to five years and by most venture capitalists to be sometime just after our sun is scheduled to go nova. The true date undoubtedly lies somewhere between these estimates.

One of the challenges with such requests is to provide a simple way for the operator to make the request to the robot's *dispatcher*. The robot may have hundreds or even thousands of possible destinations in its database. Picking through such lists, as shown in Figure 16.4, can be tedious and time-consuming.

There are several methods for accomplishing this sort of job selection. In the case of fixed cameras, a *destination node* can be specified for each camera's field of view. In the ideal case, the relationship between the camera being viewed and the appropriate *destination node* can be electronic and automatic. If this is not practical, then *destination nodes* can be named according to the camera number they support.

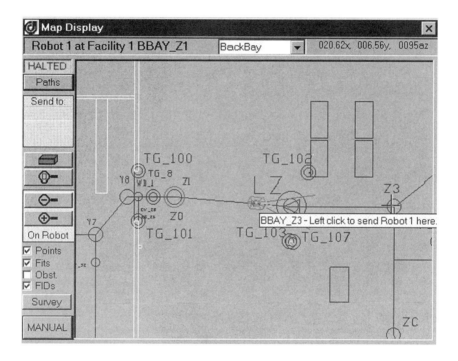

Figure 16.5. Dispatching through live maps
(Courtesy of Cybermotion, Inc.)

Another method of easing manual job selection is through live maps. In Figure 16.5, the operator has moved the cursor over the destination node Z3 on the BBAY map. To confirm that the system is capable of sending the robot to this destination, a "tool tip" has appeared telling the operator how to dispatch the robot to the Z3 node. The information that the dispatcher program needed to accomplish this was passed to it from the graphical programming environment.

Exceptions

As a robot goes about its work, there will be times when things cannot be accomplished as scheduled or planned. These incidents are often referred to as *exceptions*. It is critical that the system handle as many of these *exceptions* as possible automatically, as long as safety is not jeopardized. There are avoidable and unavoidable kinds of *exceptions*, so it is useful to consider the possible sources of such system challenges.

Programming exceptions

Most systems will undergo a shakedown period immediately following their initial installation. During this period, path programs or system parameters may be trimmed and modified for the most efficient and safe operation. Ideally, this period should be as brief and uneventful as possible. There are two reasons why this is important: first, extended debugging periods add to the system's overall cost; and second, the customer will be forming the first (and lasting) impressions of the system during this time.

We have already discussed the importance of providing expert assistance to the path programmer to assure that programs are safe even before they are tested. *An expensive robot wandering through a high-value environment is not the optimal syntax checker!*

After the shakedown period, there should be no *exceptions* as long as the environment remains unchanged. But here we have an inherent conflict; if the system is able to handle most would-be exceptions automatically, it may mask its deficiencies during the shakedown period. For this and other reasons, logging and runtime diagnostics become absolutely essential. So important are these requirements that they will be discussed in detail in a subsequent chapter.

Clutter exceptions

By far the largest source of exceptions in a well-designed and installed system will be clutter. So adept at circumnavigation are humans that we do not readily appreciate the effect of clutter on a robotic system. Clutter can physically obstruct the robot or it can interfere with its ability to see its navigational references. I have seen many forms of clutter along robot paths, ranging from an entire library of books to dozens of tiny plastic reinforcing corners taken from cardboard cartons.

And again there is a price to be paid for machine competence—the better a robot is at compensating for clutter, the less pressure will be put on the prevention of clutter, and thus the more accomplished the staff will become at creating these navigational challenges. In some cases, this can reach amazing levels.

Flashback...

I am reminded of an interesting test installation we supported in the early 1990s. The test was designed to determine the reliability of the system under ideal conditions, and to measure the number of circuits the robot could make in an 8-hour period. The installation was in the basement of a government building, and consisted of a simple "T" shaped course formed by two fairly long hallways. The system was run after hours, and in the morning the log files were checked to determine how well it ran.

For the first week or so the system performed reasonable well. There were some areas where the walls jogged in and out, and since we were inexperienced in how best to handle this sort of situation, we made a few changes to the initial programs before declaring the system debugged. Shortly after we declared victory things began to go mysteriously wrong.

The log files began to show that the robot had performed repeated circumnavigation, and in many cases had been unable to successfully extricate itself and had shut down. At that time we had no automatic ability to take routes out of service, so a failure to circumnavigate would leave the system halted in place. In any event, there was only one route to each place, and thus no possibility of rerouting. Significantly, the robot would be found in places where there was no obvious clutter, leading us to believe that it had become disoriented navigationally and tried to find a path through a wall.

Over the next week we increased the amount of information we logged, and we even created an "incident report." This report would serve the same function as a flight data recorder, and allowed us to see the robot's final ordeal in great detail, including parameters such as its uncertainty. To our great surprise, the navigation looked solid. We next began to question the sonar sensors, but no fault could be found there either. Any time someone stayed to observe the system, it performed beautifully (of course).

Finally, out of frustration, we placed a video recorder inside the robot. We connected the recorder to a camera that had been on the robot all along, but which had not been operational. Thus, from the outside the robot appeared unchanged.

After the next shift we collected and viewed the tape. The first hour or so was uneventful, and then the night shift security guard appeared. On the first pass he simply blocked the robot for a short period, causing it to stop and begin a circumnavigation maneuver. He allowed the robot to escape and continue its rounds, but he was not done tormenting it.

By the time the robot returned to the same area, the guard had emptied nearby offices of chairs and trashcans and had constructed an obstacle course for the robot. On each successive loop the course became more challenging until the robot could not find a way through. At this point, the guard put away all the furniture and went off to find new diversions!

In a properly monitored system the robot would have brought this situation to the attention of the console and the game would have been short-lived. This is an excellent example of why robots will continue to need supervision.

Hardware exceptions

Hardware exceptions can be of varying degrees of seriousness; for example, a servo may exceed its torque limit due to a low-lying obstacle on the floor like a crumpled throw rug or the box corners discussed earlier[2]. While the servo has plenty of power to overcome such an obstacle, the unexpected torque requirement may cause an exception. The SR-3, for example, is capable of learning the maximum drive and steer power required for each path leg. In this case the robot might be safely *resumed*, but it is important that an operator make this decision[3].

Other hardware exceptions can result from temporary or permanent malfunctions of sensor or drive systems. In each case it is a good idea for someone to assure that the robot is not in a dangerous situation. In the worse case, the operator may need to drive the robot back to its charger or ask someone to "tether" it back locally.

Operator assists

When the automatic exception handling of the system either fails or determines that the operator needs to make the decision to go on, it stops the robot and requests the operator to *assist*.

A security robot might typically perform 110 to 180 jobs over an 8-hour shift. As a benchmark, we have found that when the robot requires more than one assist per 300 jobs, the operators will tend to become annoyed. In later installations we observed assist levels averaging one per 1350 jobs. The assist level is an excellent indicator of how well the robot is performing.

---

[2] Interestingly, since the robot in that case was using sonar illumination of the floor, the tiny corners made excellent sonar retroreflectors and the robot was able to avoid running over them. However, had they been encountered on a carpeted surface in narrower confines, this might not have been the case.

[3] Because of such interlocks, there have been no incidences of Cybermotion robots causing bodily harm or significant property damage in almost 20 years of service.

Exception decision making

In dispatched robot systems, exception decision-making will be distributed between the robot, the dispatching computer, and the operator. The robot will normally be able to handle most simple cases of clutter-induced exceptions, eliciting advice from the central station only after its own attempts at handling the situation have failed. The robot's aggressiveness in attempting to handle an exception may be set high in one area and low in another, depending on the environmental dangers present. For example, aggressive circumnavigation may not be appropriate in a museum full of Ming Dynasty pottery.

Onboard decisions

Normally exceptions will come in two types: *fatal* and *nonfatal*. A *fatal exception*, such as a servo stall, will immediately halt the robot and will require an operator acknowledge that it is safe to resume before the robot attempts further movement.

A servo limit warning is an example of a *nonfatal exception*. A servo limit warning will normally precede the stall condition, and can be handled by the robot itself. For example, a robot that begins to approach its stall limit on a servo should, at the very least, stop accelerating. Such reflexive behaviors can greatly decrease the number of fatal exceptions that the system will experience.

Host decisions

If a robot becomes blocked and cannot circumnavigate the obstruction, it is possible that the system will still be capable of handling the exception without bothering the operator. In these cases, after halting, the robot will indicate its problem to the dispatching computer. The nature of the problem is normally communicated through the robot's status.

For a blockage, the dispatching computer will normally suspend the offending path, so that all robots will avoid using it for a predetermined time. In the case of map-interpreting systems, the *danger object*[4] found by the robot will be placed into central memory and will be sent to all robots that use the same map.

Once this has been done, the robot will be rerouted to the destination. If there is no alternative route, then the reaction of the dispatch computer will depend on the ap-

---

[4] See Chapter 15.

plication. For example, a security robot dispatcher may simply delete the aborted job from its *job queue*, or it may place a note in the queue that will reactivate the job if and when the suspended path comes back into service. For a material-handling robot, the decision making process may be quite a bit more complex, as it will depend on whether the robot is carrying a load, and if so, what should be done with that load.

Operator decisions

Operator decisions about exceptions should be kept to a minimum. Even so, issues that involve judgment or knowledge of things beyond the robot's reach should be referred to the operator. Most common among these situations are safety issues.

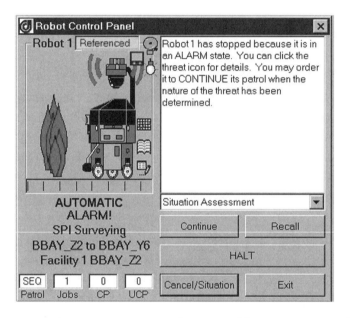

Figure 16.6. A pop-up exception panel in response to fire
(Courtesy of Cybermotion, Inc.)

Figure 16.6 demonstrates the operator display that resulted from fire detection by an SR-3 robot. Since this robot has stumbled upon this fire in the course of normal operation, and not as the result of checking out an external alarm, the situation represents a new incident and requires operator intervention.

Since such a situation will no doubt come as something of a surprise to the operator, it is important to provide as much advice as possible to aid in the proper response. This advice, which appears in the *situation assessment* window, is called an "expert," and it is spoken as the menu pops up. This is an excellent use of a text-to-speech system.

Expert assistance

Everyone who has used a modern computer has used "look-up" help and seen pop-up messages. Expert assistance is very different from these forms of help in that it immediately tells the operator everything that is important to know and nothing else. To accomplish this, an expert system compiles sentences from the status of the robot and from other available data.

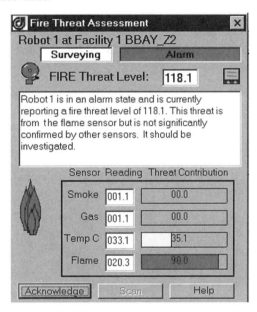

Figure 16.7. Fire threat assessment display

It is also critical that the operator be given a way to immediately access related data. In the example of Figure 16.6, there are several things the operator may want to know, including which sensors are detecting the fire, and where the robot and fire are located. At times of stress, operators tend to forget how to access this information, so simple graphical rules with ubiquitous access points are in order.

For example, in Figure 16.6 the operator could click on the flame to obtain the *threat display* shown in Figure 16.7. This display contains immediate readings from all the fire sensors, the fuzzy logic contributions of the sensors, and another expert window. The operator can also click on the "floor" under the robot in Figure 16.6 and launch a map display showing the position of both the fire and the robot.

Additionally, almost every graphic on these displays can be clicked. The small blue monitor, for example, indicates that the robot is equipped with digital video and clicking it will bring up a live image from the robot's camera. If the sensors have determined the direction to the fire, then the camera should already be pointing at the source. Clicking one of the individual sensor bar graphs will elicit a graph of the recent history of that reading, and so forth.

Status monitoring

One of the important functions of a central dispatch program is to permit the easy monitoring of the system. Text status messages do not attract the operator's attention and are not as easily interpreted as graphical images. For this reason, we chose to keep all aspects of the i-Con status display as graphical as possible.

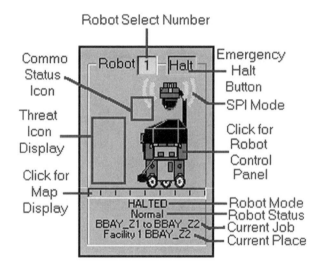

Figure 16.8. A single robot status pane

For i-Con, we chose a multidocument interface similar to that used in popular office applications. Along the top of the display a set of robot status panes is permanently visible. Each pane has regions reserved for certain types of display icons, some of which are shown in Figure 16.8. Any graphic in the pane can be clicked to elicit more information about that item or to gain manual control over it.

The color red was reserved for alarm and fault conditions, while the color yellow was reserved for warnings. We will discuss some of the rules for developing such an interface later in this chapter.

Taking control

Occasionally, it may be desirable or necessary for an operator to take control of a robot and to drive it remotely. This should only be done with the collision avoidance active for obvious safety reasons.

In addition to providing collision avoidance, sensor data can be used to interpret and slightly modify the operator commands. For example, if the operator attempts to drive diagonally into a wall, the system can deflect the steering command at a safe distance from the wall to bring the robot onto a parallel course. Such control is referred to as being *tele-reflexive*.

Driving by joystick

The traditional method for accepting operator input for remote control is through a joystick, but this adds hardware costs to the system, clutters the physical desktop, and is appropriate only when the operator is viewing live video from a forward-looking camera. When it is actually needed, the joystick will most probably be inoperable due to having been damaged or filled with a detritus of snack food glued together with spilled coffee and colas.

It is possible to eliminate the requirement for a mechanical joystick by creating a graphical substitute as shown in Figure 16.9. To direct the robot forward, the operator places the cursor (as shown by the white arrow) in front of the robot and holds the left mouse button down. To turn, the mouse is moved to one side or the other of the centerline. The speed command is taken as the distance of the cursor from the center of the robot graphic.

As the servos begin to track the commands of the "joy mouse," a vector line grows toward the command position of the cursor. If the line reaches the cursor, it means that the robot is fully obeying both the drive and steer command. If the vector does not reach the cursor, it indicates that the sensors are not allowing the robot to fully obey one or both of the velocity commands. If the cursor is moved behind the robot, then the robot will move in reverse.

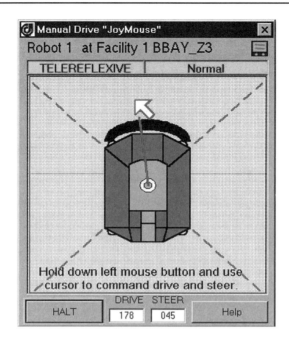

Figure 16.9. The "joy mouse" graphic substitute for a joystick

Operators found this display to be very natural and easy to use. Since it is most often used in conjunction with video from the robot's camera, the camera is automatically locked in the forward position while the *joy mouse* is open. For convenience, a blue "monitor" icon allows the video display to be opened from this form, as well as from many other forms.

Driving by map

In some cases, the operator may wish to direct the robot from a map perspective rather than from a robot perspective. For this purpose, the standard map can incorporate a manual drive mode in which the robot chases the position of the cursor on the map. I-Con also has this method of manual driving which I have personally found very useful when remotely testing new programs.

In the case of P-code programming, the robot receives its navigation cues from each path program, so if it is driven manually it will only be capable of deducing its position changes from dead reckoning. This should be quite accurate for shorter distances, but will become geometrically less accurate as the robot drives longer distances. Here, an onboard map interpreter has a distinct advantage in that it can continue to search for and process navigation references as long as the robot remains in the area defined by the map.

Some GUI rules

We have touched upon some of the ways a graphical controller can ease the operator requirements and help assure proper operator responses to various situations. Some of the rules for a good GUI (graphical user interface) are summarized here as a reference and check list.

Use appropriate graphics

It is true that a picture is worth a thousand words, but an obscure picture is worse less than nothing. I am often amazed at the quality of shareware that is available for free, or nearly for free, on the internet; yet I find that the biggest deficiency of this software is usually not in its operation, but in its operator interface. Puzzling over some of the obscure symbols that these programs use, I find myself empathizing with Egyptologists as they gazed upon hieroglyphics in the days before the discovery of the Rosetta stone. Graphics should be kept simple, bold, and appropriate.

Provide multiple paths and easy access

One of the interesting aspects of mainstream modern software is the number of different ways any one item can be accessed. If you have ever watched someone use one of your favorite programs, you have probably noticed that they know ways of navigating from feature to feature that you didn't know.

Different people prefer different methods of navigation. Some prefer graphical objects, while others like buttons and still others fancy drop-down menus. There are even a few die-hards who still prefer control keys. With modern operating systems, it takes only a few lines of code to enable an access path, so it is important to provide as many of these options as is practical.

Prioritize displays and menus

Displays should be prioritized so that the important information cannot be inadvertently obscured by less important data. Moreover, menus and lists should be prioritized according to the frequency at which they will be accessed.

I am also amazed when highly complex and well-written programs seem to have intentionally hidden paths to one or more commonly used features. Such functions should be easily available in both graphic interfaces and in drop-down menus. In menus, this poses a trade-off between the number of items in any given list and the number of nesting levels, but there is no excuse for hiding an unrelated function in an obscure menu branch. Remember that a few moments of the programmer's time can save countless man-hours for the users.

Be consistent

If clicking the camera image of a graphic display presents a control panel for the camera's pan and tilt, then clicking the lidar should present its display. If clicking a bar graph of sonar range provides a graph of the sonar echo, then clicking a bar graph for another sensor should have a similar result.

People very quickly and subconsciously learn cause-and-effect rules, but exceptions poison the process.

Eliminating clutter

One of the biggest problems for a GUI interface comes as a control program grows to have more and more features; great panels of buttons and controls can result. As the number of controls increases, the space available for displaying their functions decreases and as the captions on buttons and graphics become smaller, the function for each control becomes more difficult to impart.

It is therefore essential to minimize the number of controls displayed at any one time. If a robot is running, then there is no need to waste space with a "Start" button, and if it is halted, there is no need for a "Stop" button. Instead, one of these buttons should replace the other as appropriate.

Group controls together logically

While this rule should be common sense, it is often ignored. The trick is to think about which controls would likely be needed in response to a given display. For example, the video display shown in Figure 16.10 includes controls for the camera pan and tilt. The pan and tilt mechanism is a different subsystem on the robot than the video digitizer, but the two logically work together. At the control level, they appear to be one system.

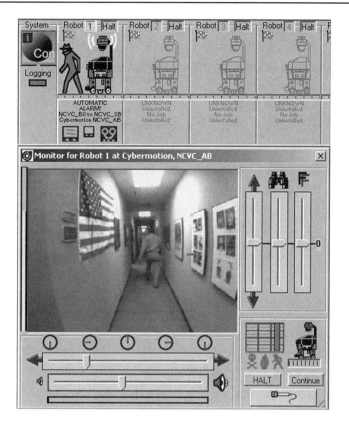

Figure 16.10. Partial i-Con screen showing status panes at top and video and camera controls on desktop at bottom
(Courtesy of Cybermotion, Inc.)

An operator very often chooses to view video because the robot's sensors are indicating an intruder. These sensors can actually direct the camera to stay on the intruder if the operator is not manually controlling the camera. To indicate that the camera is in the tracking mode, crosshairs appear at the center of the image, and a "mini-bar graph" displays the threat detection level in the bottom left of the monitor window.

Small bar graphs along the left and bottom of the image window show the camera video level and the audio level, respectively. Since these levels are measured at the robot, they serve as a diagnostic tool. For example, if the screen is black, but the video level is high, then the camera is working but there is some sort of problem from the digitizer to the base.

The operator may decide to direct the robot to resume its patrol, or to halt permanently until the situation is stabilized. Since these two commands frequently follow the viewing of video, they are included on the display as well as on the control panel shown in Figure 16.6. Finally, a "push-to-talk" button allows the operator to speak through the robot to anyone in its vicinity.

Color coding and flashing

Color coding and flashing are excellent ways of attracting attention to items of urgent interest. For these to be effective, it is imperative they not be overused. As the result of conventions used in traffic control, we naturally associate red with danger, yellow with caution, and green with all-systems-go. If these colors are used exclusively to impart such information, then a display element bearing one of these colors will be instantly recognizable as belonging to one of these three categories.

Use sound appropriately

Sound is also an important cue, but there is an even lower threshold for its overuse. If long sound sequences are generated during normal changes of state, the user will begin to look for ways to disable it. Operators who have yet to master the proper use of the power switch will quickly discover very complex sequences required to disable annoying sounds.

Likewise, synthetic speech can be very useful, but if it repeats the same routine messages repeatedly it will be ignored, disabled, or destroyed. It is therefore useful to provide options as to what level of situation will elicit speech, a short sound, or no noise at all. One solution is to link the controls of speech to the color-coding scheme. Under this approach, the system administrator will have the option of enabling speech for alarms, warnings, and advice separately.

Provide all forms of help

Besides expert systems, conventional *help files* can be useful. If these are used, it is worthwhile to link them to open displays—for example, if a help button is provided on a map display, it should open the help file to the section discussing the map and not to the beginning of the general help file. It is often appropriate to link the state of a display to the help files in such a way as to minimize searching.

Another great cue is the *tool tip*. A tool tip is text that pops up when the cursor is held over an object for a defined period. Tool tips should be provided for all visible controls so that the operator can determine the nature of the control without the need to execute it.

Finally, the cursor itself can be changed to indicate the state of a particular form. Many modern applications feature cursor symbols such as an hourglass to indicate that a form is waiting for some function to be completed. If a graphic is hot (clickable), then a cursor change and/or an appropriate tool tip should be displayed when the cursor is placed over the active area.

Robustness and context preservation

It should be obvious that a dispatching program will be used for long continuous periods and must be totally reliable. Since there will be thousands of variables, there will be an almost limitless combination of states, and it will be almost impossible to prevent an error from eventually occurring. Every function should include well thought out and tested error handling so that the application will not crash if it encounters an error.

Additionally, the program should constantly save its state and the states of the robots it serves to a level that it can be rebooted and immediately resume operations where it left off. If this is not done, the system may take a very long time to bring back online if it is restarted, intentionally or unintentionally.

Conclusions

A dispatcher and control application will be essential to most high-end mobile robotic systems. The complexity of this program often rivals the most complex commercial computer applications. If the system is to be accepted, this program must be powerful but intuitive. Above all, it must be indestructible!

CHAPTER 17

The Law of Conservation of Defects and the Art of Debugging

Perfection is a lofty and worthwhile goal, but it has never been achieved by a mortal. We will make mistakes and more of them than we care to believe. Debugging is therefore a critical part of the design, production, and installation phases. Debugging becomes more difficult as systems become more complex and therefore less deterministic.

By very nature, the programming of an autonomous robot is virtually nondeterministic. I say "virtually" because theoretically the robot should respond in a deterministic manner to any given set of inputs that occur within the same exact state of the machine, and with the same recent history. However, the permutations of these preconditions are so nearly infinite that no given set is likely to occur twice in a robot's lifetime! For this reason, of all the skills a designer can attain, arguably the most important is the skill of debugging.

Before one can become a Zen master in the art of debugging, it is essential to understand the nature of defects (bugs). Engineers have long agreed on the immutable nature of the famous "Demo Effect" and "Murphy's Law," but it has only recently been discovered that these are merely corollaries of the more encompassing *Law of Conservation of Defects*. A layman's appreciation of this law will demonstrate the basis for the art of debugging.

The law of conservation of defects

This natural law is believed to have been first expressed by engineers at Cybermotion, Inc. during the 1980s[1]. Its primary precept is:

---

[1] Blachman, Hammond, and Holland, *Recollection from discussion in engineering department men's room*, March 14, 1984.

Defects can be neither created nor destroyed, they can only be displaced. It was originally believed that the number of defects in the universe was increasing, but it has since been determined that human activity (especially technological activity) is simply attracting naturally occurring defects. Defect theory holds that as mankind destroys ecosystems and other species, the defects in those systems migrate toward human systems as governed by the rules described below. Since we cannot destroy defects, it is left to us to attempt to force them out of our systems and into someone else's, and/or to prevent their migration into our systems. Thus, it becomes extremely important to understand the dynamics at work in the mobility of defects.

1) *Sentience – Defects move and manifest themselves in such a way as to disrupt progress and to instill humility.* Although there is a great deal of controversy over the mechanism, defects undeniably exhibit what appears to be this sense of purpose. While the debate still rages as to whether defects are truly sentient, the forces affecting their mobility make them appear so. If one is tempted to doubt this rule, simply try to think of an example of a defect accelerating our progress or making us feel more self-confident.

   Several attempts to prove or disprove the sentience argument through the creation of Turing-like tests have been suggested, but these attempts have thus far been thwarted by the *stealth* and *observability* characteristics of defects as discussed below[2].

2) *Natural selection – Defects demonstrate an enormous diversity of potency and therefore are subject to natural selection.* Slow and imprudent defects are easily detected and thus forced from their hosts, leaving an evermore capable population of defects remaining.

3) *Stealth and concealment – In order to accomplish their purpose, defects attempt to avoid detection by manifesting themselves only when they can accomplish the maximum disruption.* This appears to be a natural result of the laws affecting defect mobility, but there are those who believe it is a further sign of sentience. Perhaps the most compelling argument for defect sentience comes from the observation that defects have an uncanny ability to sense our assumptions and use them to create a highly protective logical shell. The *stealth* effect also explains the attraction of defects to complex systems where they can easily find concealment.

---

[2] Research collaboration effectively ended when one of the two men's room stalls was deemed "handicapped" and thus not available for further research discussions.

4) *Observability – The likelihood of a defect manifesting itself decreases geometrically when observed by an individual capable of unmasking and correcting the defect.* This corollary explains why complex systems always go down when the resident expert is away. If a system goes down while an "expert" is present and watching, it is likely that it is because the individual is not capable of correcting the problem. This law is immutable.

Furthermore, an obvious and repeatable defect may suddenly exhibit *stealth* as the result of attempts to demonstrate it to others. Comments such as "Hey, guys, look at this!" can send a defect into instant remission, only to reappear when the additional observers have left the vicinity.

5) *Camouflage – Defects are capable of disguising themselves in such a way as to divert suspicion away from their true nature.* Those who perform maintenance on their own automobiles know that by simple intermittent manifestation, a faulty fuel pump can exhibit the exact symptoms of an ignition module failure. The symptoms will change to simulate a third possibility when the ignition module has been replaced. The profitability of the entire auto parts industry is dependent on this phenomenon.

The camouflage capability of defects is also observer dependent. To a hardware engineer, the problem will appear to be clearly a software issue, and to the software engineer it will appear to be a hardware problem. As confirmation of this principle, try to remember a conversation with a technical support hotline in which the technician suggested the problem might be in their product. QED.

6) *Criticality attraction – Defects are attracted to objects by a force that is proportional to roughly the square of the importance of the object's proper operation.* Thus, a system can be successfully demonstrated to colleagues, visiting scout troops, and the cleaning staff for thousands of hours, only to fail immediately when demonstrated to the sponsor. This characteristic is of course responsible for the famous "Demo Effect."

The well-known "Murphy's Law" has improperly been shortened in popular usage to state, "If anything can go wrong, it will go wrong." In actual fact, it is known to have originally included variance over the axis of criticality due to the *criticality attraction* and *stealth* effects of defects. The proper statement of Murphy's Law is actually "If anything can go wrong it will go wrong, and at the worst possible moment."

7) *Manifestation threshold decay – The likelihood of a defect manifesting itself increases with time.* However, once having manifested itself, a defect will tend to return to dormancy through any subsequent attempts to force its appearance by duplicating the conditions of its manifestation.

8) *Symbiosis – Defects can affect each other in such a way as to enhance their mutual concealment.* Attempts to duplicate the conditions present when one defect was manifested will sometimes induce other defects to manifest themselves in such a way as to draw the attention away from the threatened defect. Those who believe defects are sentient use this apparent cooperative behavior as another pillar of their argument.

9) *Polarization – Defects are attracted to a moving object as it passes them in space, but having become polarized by movement through the earth's magnetic field, they are subsequently repelled if the object reverses direction.* Thus, if a complex system is shipped from east to west, it will tend to pick up defects in transit[3], but these defects will disappear when the object moves back eastward. This effect explains why a system will fail completely when sent to a trade show, but work flawlessly when returned to the factory for repair. The further the system is shipped, the more complex it is, and the higher the shipping cost, the more defects it will attract.

10) *Resonance – The potency of a defect is positively influenced by the presence of other defects.* As the number of defects in a system reaches criticality, it becomes virtually impossible to isolate them and force them from the system. This effect explains such diverse phenomena as the Federal government and Windows 98®.

Defects are dislodged from an object by applying energy to the process of uncovering their nature. Since defects are repelled by the act of being observed, they can be forced from systems if the effort is sufficiently intense and focused over an adequate period of time. The energy required to dislodge a defect bears a direct relationship to the *potency, stealth,* and *resonance* of the defect, and an inverse relationship to the skill and perseverance of the debugger.

---

[3] Prior to the discovery of the *Law of Conservation of Defects*, it was widely believed that defects were the manifestation of discrete matter and not the result of complex wave phenomena we now know them to be. Polarization effects were therefore explained by "Grivet Theory," which held that there existed East Bound and West Bound Grivets. Today it is amusing to think that such a fanciful explanation could have gained the widespread acceptance that it did.

The art of debugging

Now let's put all this theory to practical use. There are defects in design, and defects in manufacture, defects in mechanical systems, and defects in software. Defects are intensely attracted to autonomous vehicles because there are so many places to hide! It is often hard to determine if a bug is the result of a long hidden design flaw, or of an assembly or material error, or a combination of these. Worse yet, through symbiosis, defects in mechanical systems will interact with defects in software to make isolation of either extremely challenging. In all cases, the debugging process is basically the same.

The most important part of debugging any system will have occurred before the first bug is observed. In software systems, this means that the programmer has designed every function fully expecting it to require debugging! In mechanical systems, this means that each part has been measured and its dimensions checked and documented before it is assembled into the next higher assembly. To think that one's design or program is bug-free and doesn't require this extra effort is an **assumption** that will attract defects like flies to sugar.

Error handling, error reporting, and logging should be built in at every reasonable level of the code. When a defect is later detected, the presence of such data can make it possible to filter out many incorrect hypotheses without the need to even test them. One can think of these windows into the system as repelling defects because they hate to be observed!

It is crucial to point out at this point, however, that error handling and diagnostics code is often a refuge for defects because it is so often written without being well tested. When a defect occurs in the mainline code, the errors in the diagnostic code will eagerly act to camouflage it through *symbiosis*. **All error handling, logging, and diagnostics code should be as thoroughly tested as mainline code.** Assuming that it will never actually be needed will be another irresistible attractant to defects.

Because of the *Resonance* and *Symbiosis* effects, it is crucial that defects be eliminated before they accumulate. The most effective way to accomplish this is to **test and debug components, modules, or other elements of a design as they are created**. Since these components may require input from other objects in the design, testing them may require writing simulation code or including BITE (built-in test equipment). The effort required to do this is *never wasted*, especially if it is done in such a way that it becomes a permanent part of the design.

The SWAG method

Now let's look at some actual debugging methods. Almost everyone is familiar with the basic SWAG (super wild ass guess) method of debugging. Like a wild ass, this method often leads one about on erratic paths that may never converge on the solution. Simply stated, this process consists of the three steps shown in Figure 17.1. The second step is the SWAG. The more information that we have about the occurrence of the bug, the more likely that our SWAG will be accurate.

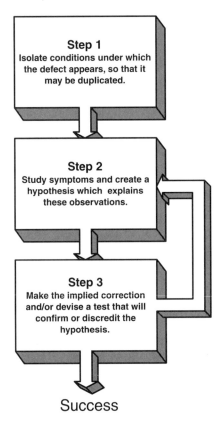

Step 1
Isolate conditions under which the defect appears, so that it may be duplicated.

Step 2
Study symptoms and create a hypothesis which explains these observations.

Step 3
Make the implied correction and/or devise a test that will confirm or discredit the hypothesis.

Success

Figure 17.1. SWAG debugging

The SWAG method works best on bugs that occur in relatively small systems or programs. Therefore, in complex systems and programs the SWAG method may be most useful when other methods have reduced the possibilities to a reasonably small subset of the whole.

If the result of step three is negative, then steps two and three are repeated until no more hypotheses can be imagined. This basic approach will quickly route the more obvious, low-potency defects, but it is a dangerous approach.

The weakness of the SWAG method is that the hypothesis generated in step two is very subject to our prejudices and assumptions. If these are wrong, we quickly find ourselves in an endless and frustrating loop—remember that defects have a knack for using our assumptions against us. The SWAG method is almost useless against multiple bugs, not merely as a result of the *symbiosis* effect, but because we usually begin with the *assumption* that one bug is responsible for all or most of the symptoms of the problem.

There is a second danger in the SWAG approach to debugging. We may make changes to test a particularly compelling SWAG, and then not remove these changes when they proved not to correct the bug. We tend not to remove them because they seemed very reasonable. These changes add a new dynamic to the debugging cycle that may create a hole in the logical net we are attempting to build around the defect! **Never keep any change made during the SWAG cycle unless it can be proven to correct at least part of the problem**.

Unfortunately, like germs that have become tolerant of antibiotics, many defects have developed the characteristics outlined earlier in this chapter in order to thwart this simple approach. Among the most effective defensive strategies are those of low *observability* and *stealth*. When a defect occurs so seldom that it is impossible to know for certain if the results of any action have indeed corrected the problem, the time required to test a given hypothesis approaches infinity. This is the kind of defect that makes error logging and exception reporting so essential.

How many times have we all muttered the famous (but universally incorrect) lines "I have tried everything" or "This is impossible"? The fact is that there is a solution, and we have simply missed it.

The divide-and-conquer method

This method is more methodical than the SWAG method, and is generally much superior. In the divide-and-conquer method, we try to find some place at which we can divide the system into big chunks and test these pieces separately. When the piece containing the defect is determined, then subdivisions within that part may be performed, and so forth until the bug is cornered in such a small area that it can no longer hide.

In hardware debugging, divide-and-conquer is usually accomplished by replacing components with known good ones. As the defective assembly is isolated, substitution can be made to its components, and so on until a cost-effective solution is realized. The cost of electronic circuit boards is now sufficiently low that in most cases it is not economically feasible to repair them. This is especially true since there are fewer and fewer easily replaceable components.

There are two basic ways to divide-and-conquer complex software problems. The first method divides the system into two pieces, and places inputs into these blocks and tests their outputs. Unfortunately, many systems on a mobile robot are so dynamic that it may be difficult or impossible to exercise the pieces statically. In this case, division is done by observing data at boundaries between sections of the system in order to determine where the error is occurring. When debugging software, the presence of public blackboard memory can be essential to this process. For example, consider the simplified block diagram of a lidar navigation processor in Figure 17.2.

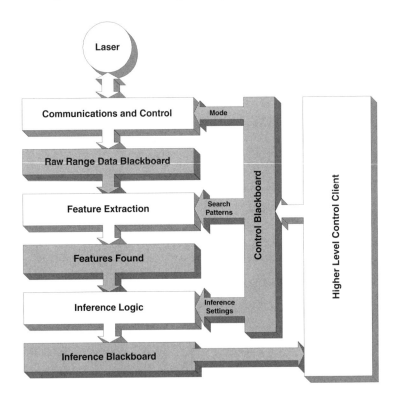

Figure 17.2. A lidar processor

Notice that not only are intermediate results available in public blackboard memory, but so are the control parameters set by the higher-level computer. Unfortunately, the raw range data and the results extracted from it will be in numeric form that is difficult for a human observer to evaluate. For this reason, diagnostic displays should be created as the design evolves to allow a technical observer to see the range and feature data in graphical form. It is also possible to write the control blackboard parameters to place the system into operation for testing, even if the real higher-level control does not exist.

Figure 17.3 demonstrates how a graphical diagnostics display can access the data available from the blackboard of Figure 17.2 and display it in a way that the user can immediately comprehend. Note that this single form has several different diagnostic displays besides the range display shown.

The bar graphs at the lower left of the form indicate the quality of the navigation results being returned for two different types of lidar navigation. The small dialog box in the center of the bottom of the form is the text output of a simple expert system. The message displayed is created from an expert *tree* that generates a critical assessment of the system's current state. Such experts can be very simple or quite complex in nature.

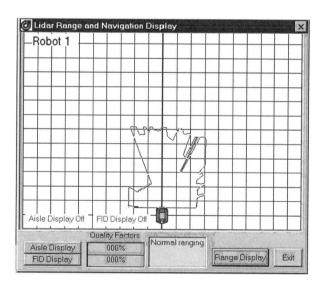

Figure 17.3. Lidar diagnostics display
(Courtesy of Cybermotion, Inc.)

If we encounter a very rare transient bug in the operation of this system, it may be very difficult to see it even with the best tools. In such an instance, we need to freeze a snapshot of certain blackboard values for later consideration. The data can then be written out to an exception file or even sent to the software provider via email. Microsoft began doing exactly this beginning with the Windows® XP operating system.

Component verification method

In the presence of multiple faults, the divide-and-conquer method may not find the problems, because it never fixes all of the bugs at one time. The component verification method is a last resort for such cases. One can think of the verification method as being the opposite of the divide-and-conquer method, in that it verifies that components are fault-free instead of searching for bad components.

Basically, this method uses a working system as a test set, by replacing its working modules with ones whose state is not known. If the system stops working, we have our bug cornered. This method is only useful in chasing down bugs when a working system is available, and therefore is of no real use in debugging new designs. ***The risk in using the component verification method is that defects may damage the good system.***

Types of bugs

Defects (or bugs) can be loosely classified according to the characteristics that they most depend on for defense against detection. The following types of bugs are the most virulent and troublesome known to man.

The assumption bug

Assume is spelled ASS-U-ME. The fundamental law of debugging is to make no assumption, and yet how often we ignore this rule. Remember that defects seek the shelter of our assumptions. The symptoms of the assumption bug are statements (or thoughts), such as:

"Power must be ok, the LED is lit."

"That is never a problem with this model."

"I checked that yesterday and it wasn't the problem."

"It didn't have any problem navigating the other hallway, and it is identical."

> **Flashback...**
>
> The hardest debug problem I can remember on a printed circuit board was caused by my assumption, "If there is a short between traces, it has to be between adjacent traces." It turned out that that simple little single-sided board had a short that apparently was in the material of the board from the manufacturer. The result was conductivity between two traces that were over an inch apart and which had other (unconnected) traces between them. The bug essentially defeated us since by the time we discovered it the board was fatally scarred by cut traces and lifted pads. We mounted the carcass of the board to the bulletin board along with a continuity tester for the amusement of visiting engineers.

Don't assume anything! Measure the power supply, and even look at it with an oscilloscope! Check that part that has never been bad before and recheck the items you checked yesterday.

The misdirection bug

This bug uses the *camouflage* effect by exhibiting symptoms that point an accusing finger at almost any component except the one at fault. Misdirection bugs become more common as software systems become multithreaded. For example, the failure to assure that all bytes of a variable will be completely stored before any task can use the variable may exhibit symptoms more like a mathematical mistake or communications error. To make matters worse, this defect will occur randomly but very infrequently.

The best bet for defeating the misdirection bug is to practice *divide-and-conquer techniques* and to suspend rational thought. Simply ignore what "couldn't be" and worry only about what is. Remember, if rational thought was of any use against this bug you would have solved the problem with a SWAG. Rational thought should be used only later when you need to explain the cure to someone important. Many experienced software engineers have raised the process of suspending rational thought during debugging to a high art form.

The symbiotic multibug

These bugs evolved as a survival response to the technique of divide-and-conquer previously mentioned. They take advantage of the *symbiosis* effect in that two or more discrete defects cooperate by appearing to be a single problem. This phenomenon happens more frequently than statistically expected because we *assume* there is only one defect and the multiple defects seek the shelter of this assumption!

273

In the worst case, there are two or more bugs situated on opposite sides of any reasonable divide-and-conquer line you might pick. A simple example is found in a string of series type Christmas lights when two bulbs burn out at once. Move your spare bulb around as much as you like and you will never find the culprit. If the spare bulb is bad, you are witnessing a truly awesome combination of the *assumption*, *misdirection*, *resonance*, and *symbiotic* bug behaviors. In fact, these bugs usually depend on the assumption that they couldn't happen as a vital part of their defense mechanism.

Bugs in test equipment are very prone to develop *symbiotic* relationships with bugs in the equipment under test! A test probe will often go open circuit at the instant that it would otherwise have exposed the problem. If there appears to be no signal, move your oscilloscope probe to a known signal source and verify it is still working. Test your multimeter before using it to measure continuity or voltage in the malfunctioning system.

The first strategy for dealing with a symbiotic multibug is to break your tests into ever-smaller granularity so that a given test is unlikely to be affected by more than one defect.

Sometimes the only way to fix a multibug is the *component verification method*.

Summary

In summation, if you find yourself making no progress on a bug, here are a few hints:

1. Identify your assumptions and verify they are true.

2. Check the three things you are most sure aren't the problem.

3. Divide-and-conquer.

4. Substitute parts into a working system.

If none of these suggestions work, my final solution is to smash the system with a heavy object. This generally ends the continued waste of time and brings great mental relief. But be careful, because the defects are prone to migrate to nearby objects as soon as the proper operation of their host is no longer of importance. Both of the author's front teeth are capped as the result of such defect migration.

CHAPTER 18

"What the Hell Happened?"

Like artists, good software designers have an intense creative drive, and there is no field of software development that is more exciting and gratifying than mobile robotics. As a result, robotic software designers tend to work long and intense hours during periods of development. Unfortunately, this motivation diminishes rapidly when the challenging tasks are completed and it is time to add mundane things like debugging, reporting and logging software. This is a primary reason why so many promising robotic projects end up failing in their first encounters with the real world.

Information is the most powerful of commodities; to discard it is a mortal sin.

This principle is not limited to robotics or even software systems; history is rife with examples of the enormous consequences of ignoring the importance of information. In fact, history itself is man's attempt to understand the present and predict the future by recording the past, and its entire course has been affected by who had and who lacked information.

One brilliant example is the battle of Midway Island, the pivotal battle of World War II in the Pacific. As a result of the U.S. having broken the Japanese JN25 naval code, the American fleet commander knew that the Japanese were going to attack Midway. Although the American fleet was much smaller than the Japanese fleet, the American commander decided to attempt to ambush the Japanese carriers in the minutes when they would be most vulnerable as their planes were returning from their "surprise" strike on the island.

But because of poor reporting, the Americans could easily have lost their advantage and the battle. The first American attack formation to find the Japanese carriers (which were not where they were predicted to be from earlier reconnaissance) was a squadron of torpedo planes. Since they were not at the designated coordinates, they arrived without fighter cover or dive-bomber support, as those groups had gone to the designated coordinates. As a result, their attacks were thwarted and they were wiped out to the plane.

As Admiral Fletcher stood in his operations center listening to the excited chatter between the attacking American torpedo planes diminish ominously to static, he was quoted as saying "What happened?" to which a staff officer replied, "One of our torpedo squadrons attacked." This response was not at all what the admiral wanted and he asked again "I mean….what the hell happened? Which torpedo squadron? Where? How many carriers were there?"

While the squadron's heroic attack was probably doomed from its beginning, the immensely valuable information that the pilots knew was unnecessarily lost with them because, in the excitement of the moment, they forgot reporting discipline.[1]

As robot designers, we do not enjoy this excuse! Today, even passenger cars are equipped with systems to log the vital data preceding an accident; so to design a robot without this capability is unforgivable.

One of the great advantages of robots is their consistency. If we program robots to report, they will report dispassionately and objectively under the most adverse of conditions, and they will never slant their reports to cover their own mistakes.

An autonomous robotic system that does not include logging and reporting is ignoring the lesson of that doomed squadron, and is (in my opinion) fatally flawed. If a mobile robot is involved in an incident, or simply misbehaves, there is *almost zero likelihood that observer or operator reports will be of any use in determining the cause.* Even when a report is accurate, it will probably be discounted because of the history of the poor quality of human reporting.

Flashback…

One morning I received an email from a customer who had an SR-3 security patrol robot with the latest digital video system. The incident report attached to the email stated that

---

[1] Luck was on the side of the Americans, and another squadron chanced upon a lone Japanese destroyer that led them to the fleet. The cream of Japan's carrier pilots were wiped out when all four of the aircraft carriers were subsequently sunk, with the loss of one American carrier. The Japanese never regained the offensive after the Battle of Midway, but had not chance intervened, the reporting omission could easily have cost the Americans the battle and the entire course of the war would have been affected!

the robot had refused to return to its charger, despite the fact that the path was totally clear. Worse, the report went on, the robot stopped in an intersection facing away from the charger path and wasn't even trying to get home. This was a customer with well-trained and competent operators, so I immediately logged into their dispatching computer and downloaded the log file to see what happened.

This robot had only one path from its charger out to the patrol area, so it was vulnerable to any blockage of that path. The log file showed that the robot had attempted to return to its charger and had found the path obstructed. Rather than stopping at the blockage, it had executed an automatic *recall* back to the nearest destination node. There, the base station router tried to find an alternative route to the charger. Since no such route was available, the robot was parked at this node and displayed an alarm warning saying that its batteries were low and asking for the operator to *assist*.

In the log file, an entry noted that a video clip of the incident had been automatically recorded and provided a link to its filename. Following the link, I downloaded and viewed the short clip and saved the snapshot below.

A pushcart had been left partially blocking the path. Since the responding officer had seen that the robot *could* have physically gotten past the cart, he dismissed it completely in his report of the situation. What the officer didn't know was that the safety rules (in this case stand-off distance settings) of the robot would not allow it to attempt to squeeze through such a tight area.

I attached the picture to my report, along with an explanation of why the situation had occurred. The report, and its inclusion in subsequent training, eliminated any such mis-understandings in the future. As an added bonus, on my next visit I found that operators were themselves proficiently and routinely using the video clips. The longest part of the process had been typing the explanation. Gathering and analyzing the data for the report took less than ten minutes, even though I was on the other side of the continent from the customer. The economic ramifications of such capabilities cannot be overstated.

Logging

Perhaps the most mundane and yet powerful record a robot can keep is a log. So powerful is the concept that mariners have kept ship's logs for centuries. Electronic logs are typically nothing more than databases that contain the same sort of information as a mariner's log.

The log's database format can be either a flat file, or relational. Spreadsheets are specialized forms of databases, and can also be used to store logs. Using a relational database for a log file can facilitate linking it to other data, but at the expense of increasing its size by a factor of typically four or more. Additionally, while a flat file can be viewed by a text editor, a relational file must be viewed by the appropriate database application.

A flat file is nothing more than a text file in which each entry is represented by a line that is terminated by a carriage return and/or a line-feed character. Numbers are represented by strings of ASCII numbers. Within each entry (line), data fields are separated by *field delimiter* characters. In the US, these characters are most often commas, but some European countries use other conventions including semicolons. A few older structures used tab characters as delimiters.

Database applications such as Microsoft's Access can read most flat files directly without having to "import" them. If Excel is to be used to view a file, then the first line of the log should contain the captions for the columns instead of data.

If a flat file format is chosen, then one must decide if it is to have fixed-field lengths or variable field lengths. Fixed-field lengths have the advantage of enabling the random access of data during reading and search operations, since the position of each entry in the file can be calculated. Obviously, this advantage comes at the expense of wasted storage space. Furthermore, there are several kinds of logs.

Operational logging

Like a ship's log, this electronic document records every change of state that the robot experiences, and every place it travels. If this log is compiled onboard the robot, it must be on disk or in nonvolatile memory to prevent its loss if the computer is reset or crashes. If the robot is centrally dispatched, the log should be kept at the dispatching computer though it may be useful to have supporting buffers and functions onboard the robot.

A logging function is a simple piece of code to create, and along with real-time diagnostics, it should be among the *first* functions written. If it is created early in the process, then calls to the logging function can be added as the operational code is written. If it is not done early, it will be more difficult later to remember all the places where logging is appropriate.

Among the items that should always be logged are:

1. Robot identifier

2. Job identifier

3. Time and date

4. Reason for entry (timed, status change, node reached, report or clip created, operator action, and so forth)

5. Referenced state (true/false)

6. Robot mode (halt, run, manual, and so forth)

7. Status (normal, stalled, lost, e-stop, bumper strike, blocked, fire alarm, and so forth)

8. Substatus (amplifies on status)

9. Map name

10. Map revision

11. X, Y, Z position

12. Heading

13. Last node

14. Next node

15. Destination node

16. Link to incident report or video clip (if reason for entry)

17. Notes

Note that text strings should generally be avoided to minimize the size of log files. All of the above fields except the *links* and *notes* can be represented by numeric values if indexes are used to reference items instead of their names. If nodes are to be logged as indexes (numbers), it is important to include a reference to the version of the map since any index to a map's nodes may change as nodes are added and deleted. To avoid this complexity, it may be desirable to refer to map names and node names by their text strings instead, and to accept the data storage penalty. Figure 18.1 shows some of an actual SR-3 security robot log file using a fixed-field length flat file structure.

It is not generally practical to log the state of every variable that might later be of interest. That level of detail can be recorded in an *incident report* when an incident occurs. The operational log file should simply include a reference to the reports generated much as the log in the earlier story included a reference to a video clip.

Operator action logging

It is crucial to log every action taken by the dispatch operator or by anyone controlling the robot directly. This information is not only essential to understanding an incident; it helps in determining improvements needed to the operator interface, and gauging the proficiency of the operators. My personal preference is to include this information into the operational log, rather than as a separate log. The first entry in the log in Figure 18.1 shows that the operator had closed the control panel display of robot number one at 2:59 PM (14:59 in 24-hour time notation). Nothing happened between that time and the automatic start of patrol at 6:30 PM.

Application specific logging

Application specific logging can take on many forms. For a material-handling robot, it may include a record of pick-ups and drop-offs. For a nuclear floor survey robot, it may contain radiation readings from its various sensors.

The SR-3 security robot, for example, keeps sensor logs that are recorded on a timed basis. Besides the ID, time, date, map, and position fields, these files include readings of gas, temperature, humidity, smoke, optical flame, ambient illumination, and other sensors. These logs are the basis for determining the source of alarms, and for creating environmental surveys. If properly analyzed, these reports can provide insights into an environment that are totally hidden from the human observer.

Flashback...

One of the earliest examples I can remember of gaining important insights from mass volumes of sensor data occurred shortly after we had equipped our robots with gas sensors. The installation in question began experiencing sporadic alarms from high carbon monoxide levels. The alarms occurred at what first appeared to be random intervals and positions. Sometimes there would be rashes of these spikes, and at other times, there would be none for several days.

First, we considered the time of day and day of the week. The first observation was that none of the spikes occurred on Saturdays, Sundays or holidays. When we plotted the spikes against the time of day, almost all events fell into one of three time windows that

Robot	Job No.	Date	Time	Map	X-Pos	Y-Pos	Mode	Event	Type	Cat.	Stat.	Description	Reserved
01	053308	04/22/2003	14:59:48	B13	014925	-01598	002	043	016	005	000	Exit Control Panel	00
01	053308	04/22/2003	18:30:00	B13	014925	-01598	002	043	016	006	000	Autostart Countdown.	00
01	053308	04/22/2003	18:30:32	B13	014925	-01598	002	043	016	005	000	Autostart sent ok.	00
01	053308	04/22/2003	18:30:33	B13	014925	-01598	002	035	003	003	015	Patroling	00
01	053308	04/22/2003	18:30:45	B13	014715	-01598	000	051	003	003	015	Standing By	00
01	053309	04/22/2003	18:30:45	B13	014715	-01598	000	051	009	001	001	JOB END at B13_PO	00
01	053309	04/22/2003	18:30:51	B13	014715	-01598	002	000	008	001	001	B13_PO to B13_SB	00
01	053309	04/22/2003	18:30:51	B13	014715	-01598	002	000	003	003	015	Normal	00
01	053309	04/22/2003	18:30:52	B13	014715	-01598	002	100	003	003	014	COMM OUT	00
01	053309	04/22/2003	18:30:53	B13	014715	-01598	002	000	003	003	015	Normal	00
01	053309	04/22/2003	18:30:53	B13	014715	-01598	002	100	003	003	014	COMM OUT	00
01	053309	04/22/2003	18:30:53	B13	014715	-01598	002	000	003	003	015	Normal	00
01	053309	04/22/2003	18:30:53	B13	014657	-01599	002	029	003	003	015	Looking at an Object	00
01	053309	04/22/2003	18:30:58	B13	014351	-01598	002	035	003	003	015	Patroling	00
01	053309	04/22/2003	18:30:58	B13	014306	-01598	002	035	017	003	001	Robot at B13_P2	00
01	053309	04/22/2003	18:31:13	B13	014306	-01719	002	029	003	003	015	Looking at an Object	00
01	053309	04/22/2003	18:31:21	B13	013920	-02193	002	035	003	003	015	Patroling	00
01	053309	04/22/2003	18:31:21	B13	013920	-02193	002	035	017	003	001	Robot at B13_AP	00
01	053309	04/22/2003	18:31:58	B13	008121	-02319	002	035	017	003	001	Robot at B13_AB	00
01	053309	04/22/2003	18:32:14	B13	008213	-04732	002	035	017	003	001	Robot at B13_BC	00

Figure 18.1. Partial flat file log as viewed with a text editor
(Courtesy of Cybermotion, Inc.)

were roughly from 7:30 AM to 8:30 AM, from 4:30 PM to 5:15 PM, and from 8:30 PM to 9:30 PM. The last group occurred only on Fridays, and was quickly identified as being caused by the cleaning crew who used propane-powered floor polishing equipment[2]. The coincidence of the remaining two groups to the arrival and departure of employees then became obvious.

A subsequent investigation determined the cause. In order to increase the available parking for the building, a garden area along the side of the building had been removed and paved over. The building's architects had placed the garden strategically to prevent parking in front of the building's main air intake. When spouses of employees dropped them off or picked them up, they would often leave their engines running in front of the duct. In some cases, the drivers would even back into the spaces literally injecting the exhaust into the air intake. The parking spaces were subsequently marked "off limits" and the problem went away, and a health hazard was eliminated!

In the case of survey robots, producing sensor files is the primary purpose of the system. Even so, neither a sensor log nor an operational log is of much value unless there is a way to find the information that is buried in the data.

Data mining using relational techniques

As a benchmark, a single night shift for an SR-3 will typically generate a log file 200 to 300 Kbytes in size. Searching this file with a text editor is time consuming and tedious to say the least. Obviously, such inefficiencies will adversely affect the bottom line of a robot's ROI (return on investment).

Robots are people force multipliers, and to make this multiplying effect compelling, every effort should be made to minimize the human effort required to install, operate, monitor, and maintain them.

Once log files are created, the data they contain may be searched or *mined* for the information of interest. Many types of information can be gleaned from such files, and it is unlikely that all of these possibilities will be conceived when the system is designed. Figure 18.2 is a powerful (though somewhat dated) data-mining application. When an operational log file is loaded to the application, it scans the file and provides the synopsis shown.

---

[2] One recurring problem detected by our robots was the disregard for proper ventilation during the operation of gas-powered vehicles. Since cleaning occurs late at night when the fresh air intakes are commonly shut down, the crews are often subjected to dangerous levels of carbon monoxide.

Data is not in itself information; it is the raw material from which information can be extracted.

The row of numbers near the bottom of the screen in Figure 18.2 contains information that allows the user to instantly determine whether the system is running normally or whether a situation needs attention. The key counts are:

- Jobs – The number of jobs performed during the shift. If this number is low, the robot is not operating efficiently.

- Assists – The number of times the console operator was required to assist the robot or to determine if conditions were safe for it to continue on its own. The ratio of assists-to-jobs is an important indicator of the quality of operations.

- Com – The number of times communications with the robot was interrupted or delayed. Short "hits" occur when a robot moves from using one access point to another, but longer outages indicate weak coverage or interference.

- Block – The number of times the robot could not circumnavigate around an obstruction in the path.

- Recall – The number of times the robot automatically recalled itself from a blockage and took the path out of service.

- Circum – The number of times the robot successfully circumnavigated an obstruction.

- NavEr – The number of times that a navigation agent failed to detect a feature it was expecting to use. This is often caused by clutter in the environment, and most navigation errors have no visible affect on operation unless they persist.

- Air – The number of air quality alarms that occurred during the shift.

- Fire – The number of fire alarms that occurred during the shift.

- Intrus – The number of intrusion alarms that occurred during the shift.

The slider near the middle of the screen can be used to manually scan through the data fields. The data for the log entry at the slider position in the file is displayed in the fields directly below the slider. Although the entire file can be accessed one entry at a time using the slider, it is normally used only to explore events that occur in the immediate vicinity of entries of interest. Manually scanning through the 1748 records in this example would take a huge amount of time, so relational methods must be used to quickly access the specific data.

We can see in an instant that this robot is operating very efficiently except that it has had to circumnavigate an obstacle about one job in every six, which is a fairly high ratio. Combined with the somewhat elevated navigation error count, we know immediately that there is some clutter in the environment. This is not uncommon for a warehouse. By plotting the places where the circumnavigation occurred, and where the navigation errors occurred, we can find the worst of the clutter. Better yet, a circumnavigation maneuver, like an alarm, will cause the base station computer to save a video clip that we can view.

Extracting simple text reports from logs

If the Eventlog application shown in Figure 18.2 indicates that a log file contains information worthy of further analysis, there are additional levels of report and graph generation available. Clicking the "Make Report" button on the Eventlog screen generates a report file (Figure 18.3) that can be read with a text editor. This report summarizes the data contained in the log file, listing details about events of interest.

Turning data into information sometimes requires displaying it in a way that is easily absorbed by the human mind. Since a great deal of our gray mater is dedicated to visual processing, information is most easily conveyed by graphic displays.

Once a report file is generated, two related files can be created by Eventlog; an X, Y, Map file (.xym), and a Time Graph spreadsheet file (.xls). These two files allow the data in the report to be displayed by position or by time. The map file can be read by the graphical programming interface (in this case PathCAD) to display data on the robot's map. The Time Graph can be read by Microsoft Excel and can then be used to generate a wide variety of graphic displays with respect to time.

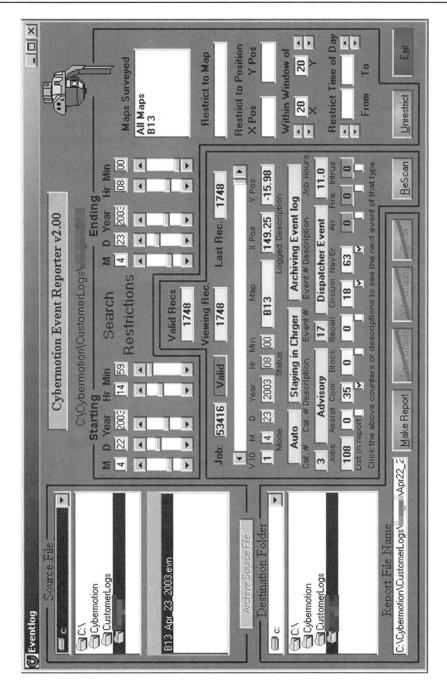

Figure 18.2. A log file data-mining application
(Courtesy of Cybermotion, Inc.)

```
Event Report Generated by Cybermotion Event Reporter v2.00.
Reporting period:
Beginning  22 Apr 2003 at 14:59
and Ending 23 Apr 2003 at 8:00

Data was not restricted by position.
This report is for the C64 map.
The graph file for this report is
C:\Cybermotion\CustomerLogs\BX2XX\Apr22_2003Ev_0.dat.

There were 1748 Valid Events reported.
Total operating hours: 11.0

                              Total for
                              Period
          Jobs Performed:         108
          Operator Assists:         0
          Communication Timeout:   35
          Navigation blockages:     0
          Path Recalls:             0
          Circumnavigations:       18
          Navigation Alerts:       63
          Air Quality Alarms:       0
          Fire Alarms:              0
          *Intrusion Alarms:        0

  * Note: Intrusion Alarms may include routine surveillance actions.


Data communication timeouts occurred as follows:

    Date        Time    X Pos   Y Pos   Map          During Job
  ---------------------------------------------------------------------
    04/22/2003  18:30   147.15  -15.98  C64          C64_PO to C64_SB
    04/22/2003  18:30   147.15  -15.98  C64          C64_PO to C64_SB
    04/22/2003  18:51   -92.97  -65.59  C64          C64_QS to C64_ES
    04/22/2003  18:52   -92.97  -65.59  C64          C64_ES to C64_DS
    04/22/2003  18:59  -107.78  -96.49  C64          C64_DS to C64_HS
    04/22/2003  19:15    35.52 -177.26  C64          C64_S0 to C64_FK
    04/22/2003  19:21   -92.99  -79.36  C64          C64_FK to C64_BS
    04/22/2003  19:49    45.33  -96.55  C64          C64_ES to C64_DS
```

Figure 18.3. Partial report from operations log

(Courtesy of Cybermotion, Inc.)

Extracting map displays from logs

Map displays answer the burning question, "Where?" By displaying events on the map, clusters and distributions are instantly apparent. This same information is contained in the text report, but staring at a column of coordinates requires far more concentration before the patterns become apparent.

As an example of the usefulness of mapping data, communications *hits* (short outages), can be correlated to the map to determine places where the communications is weak, or where delays occur as the system switches from one access point (or repeater) to another. Weak spots in communications can be determined from longer outages and they can be corrected by adding or moving radios access points.

As previously discussed, the screen of the Eventlog shown in Figure 18.2 indicated a significant number of circumnavigations had occurred. When these are plotted on the warehouse map, as shown in Figure 18.4, the location of the obstructions becomes obvious.

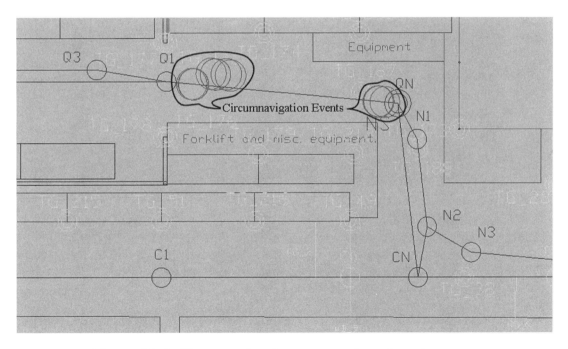

Figure 18.4. Circumnavigation events plotted against map
(Courtesy of Cybermotion, Inc.)

Extracting time graphs from logs

Time files answer another universal question, "When?" The time profile of data often yields instantaneous insights about the situation.

Flashback...

This situation I remember well because it happened less than thirty minutes ago! An operator called me about an air quality alarm from a robot that was standing in its charger and not yet patrolling. The alarm was for air quality, but it was accompanied by a fire warning because it was being produced primarily by an elevated gas reading. He indicated that an officer had investigated and smelled nothing[3]. The question was whether this was a valid reading or just a sensor out of calibration.

Upon downloading the sensor log file, I obtained the following time graph. Notice that the Gas1 sensor began to pick up an increasing reading at about 1:30 in the afternoon. The reading then continued to climb rapidly until the moment the data was transferred. The fact that the reading bobbles, rather than stair-stepping or increasing smoothly tends to strongly suggest that it is a real reading, and is being affected by air currents.

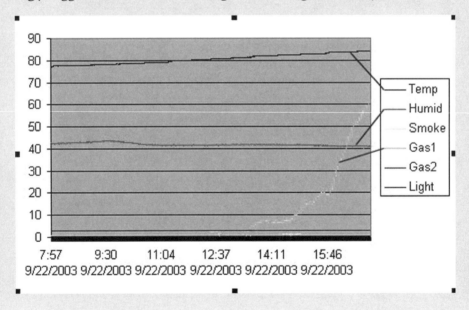

---

[3] It is interesting how people discount anything that their senses can't confirm, even though they know intellectually that there are thousands of threats that they cannot sense.

I instructed the operator to have the building ventilated and to contact the appropriate people about having the situation investigated immediately. That this example was happening just as I was extolling the virtues of logging would seem too much of a coincidence, but nonetheless it did!

Creating restricted report files from logs

An important part of data mining involves correlating data. While graphs and maps do this for some sorts of correlations, it is useful to be able to perform such correlations quantitatively and directly on the data. Spreadsheet systems can be programmed to provide this information, but it is unlikely that an operator would have the time or training to perform such an intricate task.

To accomplish simple correlations, the Eventlog data-mining application uses programmable data window restrictions. Clicking on almost any field of an entry will move that value to a restriction window. For example, if the X or Y position is clicked, restrictions are set for that X,Y position with a variable window in either direction.

If a report is subsequently created, it will average and compare all readings inside the position window to those outside of it. If there is a correlation, it will be presented in the percentage that the restricted data is above or below the unrestricted data. Once a restricted report has been generated, it can be used to create map or time displays just as discussed for unrestricted reports. Figure 18.5 shows a report comparing events that occurred between 11:00 PM and 7:00 AM to all events.

```
Event Report Generated by Cybermotion Event Reporter v2.00.
Reporting period:
Beginning  20 Mar 2003 at 18:30
and Ending 21 Mar 2003 at 8:00

Data was not restricted by position.
This report was restricted to the hours between 23:00 and 07:00.
This report is for the C64 map.
The graph file for this report is C:\Cybermotion\CustomerLogs\BX2\Mar20_2003Ev_0.dat.

There were 743 events of the 1271 total Events that met all restrictions.
Total operating hours: 11.1
```

	Total for Period	Within Report Restrictions	Percent of Total
Jobs Performed:	95	55	57.9%
Operator Assists:	0	0	
Communication Timeout:	35	17	48.6%

```
    Navigation blockages:            0              0
    Path Recalls:                    0              0
    Circumnavigations:               5              3           60.0%
    Navigation Alerts:               6              4           66.7%
    Air Quality Alarms:              0              0
    Fire Alarms:                     0              0
   *Intrusion Alarms:                0              0

* Note: Intrusion Alarms may include routine surveillance actions.

Data communication timeouts occurred as follows:

    Date      Time     X Pos   Y Pos    Map       During Job
-------------------------------------------------------------------
  03/20/2003  23:37   -39.62 -128.40   C64       C64_PO to C64_JS
  03/21/2003  00:14    81.05  -21.85   C64       C64_PO to C64_BS
  03/21/2003  00:16    92.96  -79.95   C64       C64_BS to C64_SB
  03/21/2003  00:31    45.33  -96.56   C64       C64_ES to C64_DS
                 -------- details list cut --------
```

Figure 18.5. A time-restricted operations report
(Courtesy of Cybermotion, Inc.)

Incident reporting

Logging usually records routine information, but it is not practical to continuously log every aspect of a robot's operation. Therefore, when more detail is required a special report can be generated. These documents are called *incident reports*.

Application specific incident reports

When a robot experiences an incident, it can and should generate a more detailed report about the circumstances that led up to the event. An incident can be operational in nature, such as stopping for safety reasons, or it can be application oriented, such as the report shown in Figure 18.6 for the air quality alarm we just discussed.

```
--- AIR QUALITY REPORT C:\ROBOT1LOGS\09221700.AIR ---

Robot 1 at BX2 generated this report at:
Date: 09/22/2003  Time: 17:00:23
Status: Normal
Map: BX2
Position: -0.09/-0.01
Heading: +8
Current Job: Reference BX2_PRPO
Origin Node: BX2_PR
Destination Node: BX2_PO
The Environment is unspecified.

Air Quality Threat = 142.6
AIR QUALITY ALARM
    ----------------------------
   | Sensor  | Reading | Threat |
    ----------------------------
   |  Smoke  |  000.8  | 000.0  |
    ----------------------------
   |   Gas   |  061.3  | 100.0  |
    ----------------------------
   | Temp. F |  084.0  | 042.6  |
    ----------------------------
   |  Humid. |  041.1  | 000.0  |
    ----------------------------
   | Aux. Gas|  000.3  | 000.0  |
    ----------------------------
```

Figure 18.6. Air quality alarm incident report
(Courtesy of Cybermotion, Inc.)

Operational incident reports

For an operational report, on the other hand, we are interested in different data. The report shown in Figure 18.7 was produced when the console operator halted the robot. Since this is normally done for a specific reason, it is considered an incident and causes the data from the robot to be automatically saved. Note that the creation of any incident report should be noted in the operational log of the robot.

```
--- STOPPAGE REPORT C:\ROBOT2LOGS\03312132.L42 ---
   Robot 2 at WH3  became unreferenced at:
   Date: 03/31/2003  Time: 21:32:45
   Status: Standing By
   Map: WH3
   Position: +35.82/-176.86 with certainty of +/- 0.81/0.51
   Heading: +509 with certainty of +/- 2 begrees
   Current Job: WH3_G0 to WH3_S0
   Origin Node: WH3_S0
   Destination Node: WH3_S0
   Robot was halted manually by operator.

   -------------------------------------------------
   |  Xducer     |  Left  |  Front    |  Right  |
   -------------------------------------------------
   |Front/Rear   | 05.98  | 05.95     | 06.06  |
   |Catwhisker   | 08.30  |           | 04.30  |
   |Side Coll.   | 08.65  |           | 05.25  |
   |   Wall      | 31.00  | 00.00     | 31.00  |
   -------------------------------------------------


   ---------------------------------------------------------------------------
   | Axis of Fix  |Type Fix |Last Corr |Last DLen |Last XPos |Last YPos |Curr.DLen |
   ---------------------------------------------------------------------------
   | Lateral  (x) |     FID| -000.42  | +002.48  | +035.01  | -177.33  | +000.01  |
   | Longitude(y) |     FID| +000.52  | +002.48  | +035.01  | -177.33  | +000.01  |
   | Azimuth  (a) |     FID| +0001    | +002.47  | +035.01  | -177.33  | +000.08  |
   ---------------------------------------------------------------------------

000    AVOID      000,    00375,  00130
001    TURN       000,    00512,  00000
002    *Status-Surveying
002    >SURVEY 000,    00000,  00000
003    --- End WH3_G0S0.SGV ---
003    HALT       000,    00367,  00100
   -------------------------------------------------
```

Figure 18.7. An operational incident report

(Courtesy of Cybermotion, Inc.)

As in the air quality alarm report in Figure 18.6, the operational incident report of Figure 18.7 starts with a header that describes what happened, and what the robot was doing. Following this is a table of the sonar ranges from which we can determine that the robot was over 4 feet from the nearest obstacle.

Next, basic navigation data is saved so that we can determine if the robot was operating and navigating properly. In this case, we can tell that the robot had traveled no more than .08 feet since it corrected its azimuth and .01 feet since it had corrected its position.

Finally, the program that the robot was executing is disassembled and recorded. Since the robot is executing P-code, it does not have access to the symbols used in the original programs and simply returns their numeric values. The right carat ">" indicates the instruction being executed when the robot was halted. In this case it was standing still performing a long-range sweep with its intrusion sensors when the operator decided to take it off-line, so it is unlike that this was done because of any problem with the robot itself.

Pre-event buffering

When an incident occurs, it is often useful to see the second-by-second details that led up to it, but it is usually not practical to accumulate such a quantity of data in a log file. For this purpose, we can borrow an old trick from the alarm industry and maintain a circular data store that holds a few seconds to a few minutes of data that may be of interest.

For example, the log file usually records the position of the robot only when logging events occur, such as the robot reaching a node or experiencing a status change. Saving the robot's position, heading, and other data every second can provide a rich cache of information on what happened just before an incident. When an incident report is generated, the data in the store is simply copied out into a file that can be easily displayed if needed.

In the case of i-Con, we also made this store available for the operator to review in real time. For example, a "time replay" slider at the top of the map display allows the operator to review recent events such as the movement of the robot and any intruders it observed.

Log and report management

If left unchecked, the logging and reporting functions can eventually swamp the media on which these files are being stored. Depending on operators or even field service personnel to delete or archive old files can be asking for trouble. When video files are involved, the problem can quickly become acute.

There are several ways to automatically manage this problem. The logging software can contain a programmable age limit, after which a particular type of file is erased. Alternatively, the program can contain a space allocation parameter for each type of file. When the allocated space has been filled, the program should erase the oldest file(s).

Summary

From an operational standpoint, mobile robot systems are among the most difficult software environments to predict and debug. Intrinsic software can combine with application software and environmental variables to produce enormously complex sets of conditions. It is absolutely crucial to provide every reasonable tool to help in determining what is really happening.

From an application standpoint, a big part of any automation system is the information it provides. Again, appropriate tools are essential to maintain data and to process it into useful information.

Obviously the level of logging and reporting discussed here is not needed for something like a simple area coverage robot, but even the simplest robots can collect data with very little additional cost. If the software of a robotic system does not include appropriate logging and reporting capabilities, it is very unlikely to be successfully deployed.

CHAPTER 19

The Industry,
Its Past and Its Future

Disclaimer: This chapter contains graphic descriptions of capitalistic carnage and may be disturbing to some younger robot lovers. It also contains opinions and politically insensitive observations that will be offensive to others. Reader discretion is advised.

Each spring, universities and colleges turn out large numbers of engineering students eager to find their place in the exciting field of autonomous robotics. Some have chosen technical disciplines that they feel should qualify them for such jobs, while others have attended schools specifically because they offered programs or degrees in robotics. Few are aware, at least until they begin their job hunt, how nearly nonexistent such jobs really are. As a result, the most determined of these robot engineers eventually go forth to form robotics companies in order to give their ideas life.

The bad news is that the statistics on success for such efforts are bleak. The *good news* is that the golden ring is still there to be won! It is my goal in this chapter to speak to the brave souls that are considering this quest, and if possible, give you a primer for what you will face from the business community.

The most important thing to realize is that you must be at least as creative in your business strategy as you are in your robot designs. If this challenge does not excite you, either team with someone it does excite, or don't waste your energy.

In our discussions of the technology, we have constantly observed patterns repeating themselves. This is true in business as well, so taking the time to study the past may well help us negotiate the future. To understand where we are, we must also understand where we came from and how we got here.

The history of robotics

Today, the word "robot" is used to describe a bewildering range of hardware and even software. If we accept the definition that it means a machine that can be programmed to do useful work, then the history of robotics is measured in centuries and not years. For example, Joseph Jacquard invented a textile machine that was programmed using punch cards in 1801!

The name "robot" would not be coined for another 120 years, however, when Czechoslovakian playwright Karel Capek introduced the term "robota" to describe a mechanical servant. The word "robota" in Czech translates to something between serf and slave. Soon comics, science fiction books and movies about robots began to appear, and the word *robot* was solidly established in the vernacular, if not the economy.

The man most often credited with the invention of the modern industrial robot is George Devol, who created a general-purpose, programmable robotic arm in 1954. In 1956, Devol and Joseph Engelberger formed Unimation, and a new industry was born.

Although Joseph Engelberger was an engineer, he also had a genius for promotion. The use of the term "robot" to describe these manipulator arms came only later as one of several marketing innovations. After years of attempting to sell these revolutionary devices through traditional industrial marketing channels, Dr. Engelberger appeared on Johnny Carson's "The Tonight Show" with one of the company's robots, and the reaction was totally unimaginable. Within months Unimation was being flooded with orders. In 1975, after almost two decades, Unimation experienced its first profitable year.

> ### Flashback...
>
> I have had the distinct pleasure of knowing Joe for many years. Several years ago, I attended a gala dinner in Detroit honoring his contribution to the robotics industry. In a brief chat over cocktails, I asked him about that fateful show.
>
> He said in retrospect that he was not prepared for the response that followed, and that had he known that it would spawn more competitors than customers he might well have reconsidered doing it!

A few years before, Richard Hohn at Cincinnati Milacron had also developed a robotic arm, but his was the first to be controlled by a minicomputer. Unfortunately for Unimation, the wild enthusiasm that started with "The Tonight Show" program inspired Cincinnati Milacron and attracted the attention of the management of many companies including such giants as General Electric and IBM. By 1980, the industrial robot industry was growing at a staggering rate, at least in terms of competitors and products. Unfortunately there were only three significant customers in the US, and those were: General Motors, Ford, and Chrysler.

When new entrants want to capture a piece of an emerging market, they invariably run into the patents of the pioneers like Unimation. As a result of the search for ways around the existing patents, new innovations proliferate[1]. During the heady times of the industrial robot industry, the new entrants studied the weaknesses of the hydraulic robots developed by Unimation, and new designs emerged. These configurations included all-electric robots and even pneumatic robots. Although hydraulic robots maintained a position in the heaviest applications, the electric robots quickly began to dominate the more lucrative medium and light applications.

At this point, one has to speculate as to why things went terribly wrong, but they did. The most likely cause is that in their haste to dominate the market, the competitors failed to study and understand it. This appears to have been fueled by huge and widely publicized orders from the big three automakers, and by wild predictions of future growth by industrial surveys[2].

But why did the big three automakers throw caution to the wind and embrace this new and largely untried technology so eagerly? The most likely answer comes from the fact that they saw the world in terms of management vs. labor, and robots seemed to be their perfect answer to throwing the balance in their favor[3].

---

[1] Though this may seem a bit unfair to those who sacrificed themselves to start the new industry, it is one of the strengths of our system of enterprise, at least from the consumer point-of-view. Unfortunately there are many weaknesses as well, and these will be apparent as the story continues.

[2] These surveys are produced by polling large numbers of people in the customer and provider communities. For mature industries, this process can be quite accurate because those being polled are knowledgeable and experienced. Unfortunately, in the emerging stages of a new technology or industry they tend to measure momentary enthusiasm and seldom predict the future. Unfortunately, corporate managers often use these reports as a shortcut to clearly understanding the industry and forming their own unique strategy. Additionally, the report can serve as a convenient justification for bad decisions when the stockholders begin to break out the the torches and march on the castle. The result is known as the "Lemming Effect," and there is no better example than the early years of the industrial robot industry.

[3] One may be tempted to think that investments involving large amounts of money and highly trained and experienced management would be made by an analytical and rational approach, but this is often not the case. Corporate CEOs are often influenced by power, prestige, greed, fear, and most of all, a highly competitive spirit. When these influences overshadow their rational judgment, colossal mistakes are often the result.

The first thing that went wrong was caused by the assumption that robots could simply be injected into the assembly lines to replace human workers. Unfortunately, the rigid structure of the assembly lines in those days had no contingency for a robot breaking down. While it had been rather easy to replace humans as needed, it was not so with the robots. As a result, lines often halted or vehicles were passed along that would need expensive rework later.

Worse yet, electronic communications and computing were primitive, and as a result, systems tended to be poorly integrated and monitored. And finally, the vulnerability of the robots to sabotage by threatened and disgruntled workers[4] meant that the early robots were headed for real trouble.

It is sometimes said that all new technologies go through five phases: wild enthusiasm, growing disillusionment, a search for the guilty, the punishment of the innocent, and the promotion of all nonparticipants. The industrial robotics industry slid rapidly into the latter phases. Customers began canceling orders and new orders dried to a trickle. Large American companies began to withdraw from the industry, and independent American companies fell like flies. In many ways, this was a dress rehearsal for the even more spectacular collapse of the dot-com industry that would transpire two decades later.

As American managers and financial institutions were selling their robotic houses for scrap and trying to distance themselves from the "failed" concept, the Japanese were establishing a love affair with robotics that would never waver.

In the years that followed, factories began to adopt flexible, cellular structures that better accommodated automation, and manufacturing engineers began to understand which functions were appropriate to robots and which were not. Enabling technologies in communications and computing evolved rapidly, and by 1990 the industrial robot industry was about $200 billion dollars a year, and it was dominated by over 40 powerful Japanese companies.

The American financial community referred to this transition euphemistically as a "shakeout," but in fact, it should properly be called "the Great American Robot Skedaddle[5]"! In blind panic, the United States threw away a lead that had taken decades of blood, sweat and tears to establish. The American economic system had over-allocated funding for small competitors, and left none with the ability to weather

---

[4] This factor is at least as serious with mobile robots in commercial applications.

[5] Skedaddle is a colorful term, popularized in the American Civil War. It means to run away in panic.

the first storm. Managers of large corporations, who had the resources to persist, did not have the vision or commitment to weather criticism of their investments. General Motors formed an alliance with Fanuc, and formed GM Fanuc, effectively blocking other suppliers from much of the market.

Unimation was purchased by Westinghouse, which soon broke the robot pioneer up and sold the intellectual property abroad. In the US, there was little or nothing left to show for all the investment, genius, and hoopla. The US financial community had, however, learned one indelible lesson: "Robots = bad"! The subsequent success of the Japanese in the industry had no effect whatsoever on this conclusion.

To understand why the financial managers and most importantly mainstream venture capitalists would seem to have learned nothing from this and other disasters, one must understand the business they are in. The myth is that the venture community scours the garages of the world looking for the next Microsoft or Apple, funds its early growth, and in the fullness of time, passes the strong and viable corporate child to the public equity market where it can grow to strong adulthood. In a few cases this is true, but it is not the rule; would that it was!

Instead, the venture community has evolved into the business of supplying stock certificates that will sell in the public market. Instead of being industrial leaders and visionaries, they more commonly create "me-too" companies almost out of thin air, injecting just enough capital to make them appear real, and taking them public. This is often done completely without regard to whether the company has any viability. It is only important that there is a market for the stock.

So when the incredible growth of the Internet caused a feeding frenzy for dot-com equities, the venture community obliged by cranking out companies by the score. Each business plan contained an obligatory revenue projection graph called the "hockey stick," which would turn acutely upward after just a few years of operation. At least as important is the fact that real companies more than a few years old were seldom marketable to venture capitalists because they had real histories that would get in the way of the "story."

In the hilariously irreverent comic strip Dilbert[6], the expected phenomenon behind the "hockey stick" is explained in a business plan with the single line "and then a miracle happens"! This is as good as many of the later stage dot-com business plans I have read.

---

[6] Scott Adam's Dilbert is required reading for any technology entrepreneur. It contains more insight into the realities of the technology business than any other source I have found

So what does all of this have to do with the market for mobile autonomous robots? After all, an industrial robot arm is almost unrelated to a mobile robot. They have different functions, different challenges, and different customers! The answer is that the association of robots with financial pain has yet to be forgotten (though it appears finally to be fading). This has meant the almost complete absence of second and third tier funding for mobile robotics companies.

The mobile robotics industry

Before we talk about the mobile robot industry, let's face some very unpleasant facts. The first is that in the minds of most capital managers, there is no real autonomous robotic industry at all! To understand this, consider the fact that a company must reach a sustainable level of about $30M a year in revenues to even be considered a player at the bottom tier of industry. To my knowledge, no company has yet achieved this through the sales of its autonomous robots. Even so, venture capitalists are always looking for the next technology equity fad.

There is another myth, and that is that if "you build a better mousetrap, the world will beat a path to your door." This is only true if the world believes it has an unsolved mouse problem, and if your PR and advertising campaign convince the world that you have the best solution. Convincing the world you have the best solution to a problem takes a great deal of investment. The more revolutionary a concept is, the more the investment that will be required to promote it. The task of accomplishing this in light of the capital environment is daunting, to say the least.

Flashback…

It is often said that the ties formed in adversity are among the strongest experienced by humans. At Cybermotion, the culture was forged in the furnace of adversity, and a number of jokes became embedded in the culture so deeply that they could be referenced with a single-phrase or even a word or gesture.

One such joke was about a psychiatrist who was studying why some children were pessimistic and some optimistic. Failing to find any consistent reason for the differences between his test groups, he elected to see if he could change these predispositions by subjecting the children to realities in contrast to their attitudes. To accomplish this, he placed the most pessimistic child in a room full of toys and video games, while he gave the optimistic boy only boots and placed him in a room covered knee-deep in manure from the local stock yard. After a week of leaving the boys in these rooms for 8 hours a day, he interviewed them.

The pessimistic boy had pushed all of the toys to one corner and was pouting in the other corner. When the psychiatrist asked him why, he said "If I get used to playing with them, you will just come and take them away, or they will probably break, or the batteries will run down."

After reviewing 40 hours of videotape that showed the gleefully optimistic child wading through the manure, occasionally pushing it aside with his foot, the psychiatrist asked the optimistic boy why he had remained so happy. The little boy explained excitedly that "Sir...with all this horse-crap there must be a pony in here somewhere!"

The joke became so embedded in the culture, that it was referred to constantly as a reminder that we realized we were ignoring abundant evidence that we might all be better off reconsidering our career choices!

Another joke originated spontaneously as the Vice President, Ken Kennedy, and I were leaving a high-level meeting with a large corporation about a proposed joint venture. The meeting had been exciting and full of heady comments about "good chemistry" and "win-win scenarios." But we had both seen this sort of enthusiasm dissipate rapidly once the attendants had a chance to think how it might sound that they were going to spend corporate funds to build "Robbie the Robot."

As we pulled from the parking lot, Ken looked at me and asked "What did you think?" To which I answered "They gave all the right signs" and then spontaneously offered my wrist saying "Feel my pulse!" In the end, the meeting had been a waste of time, but we had a new joke. When anyone would talk excitedly about some new business or sales prospect, one had only to offer an upturned wrist to make the point!

Early mobile robot companies

The most commercially successful examples so far have been companies at the edge of autonomous robot technology, including teleoperated bomb disposal robots, and physical path-following AGVS (Automatic Guided Vehicle Systems). But even these technical cousins of autonomous robots have failed to reach sales levels that excite the investment community.

Although the industrial (arm) robot business experienced a brief honeymoon with public investors in the early 1980s, there has been little serious public capital invested in robotics since, especially in the field of autonomous robotics. At this writing, the few public stock offerings made by autonomous robot companies have been in "penny stocks," and have netted only three to ten million dollars. Unfortunately, but predictably, this proved insufficient for these companies given the barriers that they faced.

Note to entrepreneurs: If you are new at this, you must learn to think of money in two scales. In the personal scale, a million dollars may seem like a huge sum. In the business scale, it is almost a unit of measure! A good exercise for evaluating sums at the gut-level is to divide the business dollars by 1,000 and think of that amount as a personal sum. So, to your new robotics company, a $500,000 dollar investment is like someone giving you $500 personally. Unfortunately, if you personally guarantee a loan and the bank calls it, they do not perform this division.[7]

During a visit to a Wall Street firm, I was given a chart of various technologies, and the average return each had provided to its investors. At the bottom of the list was robotics, and it was the only category that had a negative average return![8]

To date, the number of small robotics companies that have come and gone remains uncounted, but it would probably need to be stated in scientific notation if it were known. Some have lasted for decades, some for only days. Add to this the incredible number of robotic projects developed at universities and by hobbyists, and it is difficult to see the forest for the robots.

In my career, I have seen hundreds of robots, each of which represented hard work, loving labor, and often financial sacrifice. Many of these robots appeared to perform the tasks for which they were designed, and some represented world class innovation. It would seem that there really must be a pony in there somewhere! If we are to find it, we must understand the potential marketplace.

Industry segmentation for autonomous mobile robots

There are several broad market segments for autonomous mobile robots. Each of these has each been explored by numerous companies, and with numerous products. These segments include:

---

[7] Never sign your home over for a business loan. If you only need the amount of a home, you are not in a real robotics business. And if you sign it over on a larger loan, the bank will use it as a lever to have their way with you and overpower other stakeholders.

[8] This was in the 1999 time frame. It is probable that the poor performance of the robotics industry has since been eclipsed by the subsequent dot-com disaster. Unfortunately, this came at the cost of vaporizing the available venture capital almost completely for some years. For the most part, venture capitalists didn't lose their own assets, but the suckers who had been buying the inflated and over-hyped companies that they were "flipping" were simply squeezed dry. When the margin calls started, the bubble burst.

Hobby, research and gadget robots

Perhaps the easiest customer for a robot component is another robot designer, be that designer a professional or a hobbyist. Today a robot enthusiast can purchase a wide variety of components ranging from complete platforms to special powered wheels. But this market is limited and will never in itself generate the breakthrough business everyone wants to see.

A somewhat larger, but still limited, market is the gadget and toy market. Products like the Sony robotic dog have done well in this market but they tend to be brief, fad-driven successes. Since these products are not autonomous robots as such, I will not go into any depth about them here.

Home robots

From a perspective of market size, the home robot represents the Holy Grail of robot markets. If a useful and effective home robot can be produced to perform tasks such as cleaning and lawn mowing, the market potential is enormous. There are, however, significant challenges to cracking this market. If such a robot does not perform a useful task cost-effectively and reliably, it falls into the category of gadget robots and will never achieve widespread acceptance. Furthermore, compared to commercial buildings, homes tend to be fragmented spaces with different requirements, and an abundance of obstacles from furniture to pets.

The best test of a new technology or product can be made when the early customer's purchases need to be replaced. If they buy replacements, then the technology probably has "legs." If customers go back to the old way of doing things, then the technology is probably a gadget fad.

To date, dozens of robotic lawn mowers and vacuum cleaners have been created with limited market acceptance. The trade-off of cost and effectiveness is challenging, to say the least. Given the amount of power required by a typical manual vacuum cleaner, it is a significant challenge for a small, battery-powered device to accomplish any but the lightest of this work. This fact has driven most entrants in the market to opt for robotic sweepers, even though they are largely limited to hard surfaces. Yet as I write these words, engineers are working all over the world to meet this challenge. So far, Roomba™ robot from iRobot would appear to have made the most significant progress.

Some visionaries believe that the answer to the home market is a humanlike android, which would be adequately intelligent and dexterous to use the same manual tools humans do in accomplishing virtually every menial task around the house. The technical challenges of doing this are compounded by the economics. By the time this con-cept becomes technically feasible, it is likely that many of today's manual appliances will already have become robotic to one extent or another.

So, how much should a home robot cost, and what should it do? Since the labor being saved is most commonly that of the homeowner, the value of a home robot is some-what subjective. Recently, I found two examples that should bracket the current answer to the cost issue. The first was a sweeping robot being marketed through Wal-Mart for $28.95. The other was a pair of 6-foot high, futuristic-looking, remote-controlled home robots being offered in the Neiman Marcus Christmas catalog for a mere $400,000. The ideal price no doubt lies somewhere within this range.

Another problem encountered in the home market is the fact that the robot must be almost entirely self-installing and extremely simple to use. For the lawnmower applications, there are also significant safety-effectiveness trade-off issues. Any manufacturer who succeeds in the lawn-mowing market will no doubt face serious product liability issues. The first time a proud robotic lawn mower owner puts a finger under the deck to see if the blade is turning, an army of class action lawyers will spring into action in a selfless effort to save us all from the evil machines![9]

Agricultural Robots

One has only to watch a giant combine moving through an endless ocean of grain to realize that robotic guidance should be a possibility. The cost of such equipment is sufficient to easily mask the incremental cost of robotic sensors and controls. Further-more, it should be a perfect application for GPS navigation since there are no overhead obstructions. So why has this market failed to evolve?

Whether they publicize it or not, almost every large manufacturer of agricultural equip-ment has, or has had, an active program to explore this potential market. Many of these programs have roots reaching back two decades! The reason that none of these companies have attempted a robotic product launch probably has less to do with tech-

---

[9] In the United States, if you appear to have money you will be sued, usually with the intent of extorting a settlement out of you. This burden on businesses is one of the greatest weaknesses of the US system. But when this happens, take it as a sign you have become a player.

nology, and more to do with the liability issues[10]. It is also possible that the customer's potential labor savings is too small in comparison to other operating costs to make the application a compelling buy. Since no agricultural giant has yet made a serious effort to market the technology, there has been no competitive pressure for others to do so.

Yet there is another promising sector to the agricultural market, and that is picking. At present this task is done largely by migrant manual labor. Since the task requires not only navigation, but visual identification of the item to be picked, along with the development of a strategy as to how to reach it, the application requires an intelligent and dexterous machine. Even so, given the incredible cost/performance advances in computing and machine vision, it is quite likely that robotic pickers will begin to appear in significant numbers in the near future.

Manufacturing

The use of mobile robots called AGVs (Automatic Guided Vehicles) to move materials in the manufacturing sector goes back over three decades. Until recently, these vehicles were not autonomous and followed a guide path on the floor. In the late 1980s, Caterpillar developed an off-wire machine that used a scanning laser and large reflective targets to navigate.

In more recent years, an increasing number of AGV manufacturers have imbued their products with more autonomous forms of navigation. Even so, neither the older-wire followers nor new autonomous AGVs have gained the kind of market share that they had hoped for. Instead, competing methods of moving materials, such as smart conveyors, have undercut the market for AGVs. Will this change?

It is quite probable that the AGV industry has made the same mistake with its application of autonomous technology that the auto industry made in their early application of robotic arm technology. That is, they have tended to substitute autonomous technology for wire-following technology without rethinking the new possibilities this could offer.

---

[10] This application represents the kind of situation that class action attorneys dream of! First, the machines are very heavy and so any accident will be likely to cause grievous injuries, the photos of which would move any but the most callous jury. This could easily trump any rational argument as to whether the robotic system was statistically safer than a manual system! Second, these large and established manufacturers have deep pockets. These companies of course understand this, and have developed the keen senses, lightning reflexes, and risk aversion seen in any successful prey species.

While wire-following vehicles struggled even to turn around on a path, autonomous vehicles can be made to be very agile. If this capacity were combined with more intelligence, the vehicles could perform a wider range of applications. Caterpillar seemed to realize this when they launched an autonomous robotic forklift, a configuration never achievable by wire-following technology.

As a further example, one of the most promising applications is for a tow-motor robot that can automatically hook up to material-laden carts and move them from place to place. Since such carts are used extensively in many industries, the application should be appealing to many potential customers[11]. The advantage here goes far beyond labor savings. Such a system would enable the precise computer tracking of the position of carts, greatly enhancing materials tracking.

In short, autonomous robots have been quietly appearing in the manufacturing sector for many years. But the fragmentation of applications, the limited size of the overall market and the outflux of manufacturing to developing nations mean that this sector is not likely to produce the October revolution of the robots.

Service Robots

There is a large body of opinion that expects the breakout of the autonomous robot industry will occur in the service sector. The reasoning is simple; it is by far the largest sector of the economy, and it has yet to experience significant productivity improvements through automation.

The largest components of the service sector are commercial cleaning and security. To put this in perspective, consider that the annual expenditure in the US for private sector security guards is approximately ten times as large as that for all security related hardware, including alarm systems, closed-circuit TV, and access control.

And as large as the security industry is, the cleaning industry is even larger. Every day, millions of workers go forth to clean the motel rooms, offices, and public buildings that the rest of us have trashed during the previous 24 hours. So huge is this industry, that it is important to the very prosperity of the economy that it should operate at the lowest possible costs.

---

[11] In 1993, Cybermotion demonstrated this possibility by equipping its SR-2 with a hitch and pulling a cart loaded with fiber spools. Unfortunately, the limiting factor at that time was that navigation was too primitive for the unstructured target environments.

The service sector and social issues

From a customer perspective, most services are overhead and not on the profit-making side of their ledgers. This fact further increases the cost constraints on services. The tremendous pressure to keep cleaning wages low, combined with that from the agro-business community, is such that it has spawned one of the most controversial and arguably hypocritical systems since slavery, the de facto exploitation of illegal immigrants.[12]

Through a mishmash of conflicting laws, the United States and other developed countries effectively encourage undocumented aliens to enter their territories to fill the jobs most natives don't want. Whether this was an intentionally contrived system is arguable, but what is not arguable is that these workers are prevented from organizing, or even protesting, by the fear of being deported. The hapless victims of the system remain silent because, despite the miserable salaries they receive, the conditions in their native lands are far bleaker.

The problem with this system is that it is not only unfair, but also sustainable. The system depends on the host nation having an impoverished neighbor, such as the relationship between the United States and Mexico. As economic balances shift, the system will eventually collapse. More importantly, like all hard-working people, these illegal immigrants yearn for a better life for themselves and their children. As they achieve this, and their children become citizens by birth, the system requires yet more illegal immigrants. And in a time of international terrorism, this stream of illegal immigration offers a perfect conduit for terrorists.

The dependence on illegal immigrants for cheap labor is already exacting a hidden cost from the perspective of the host countries by changing the cultural, linguistic, religious, and economic demographics of their communities. The natives of these countries are beginning to understand this cost and are starting to pressure their governments to do something to stop illegal immigration. If this pressure continues, it will provide more opportunity for robotic solutions.

In many ways, the perpetuation of this system is like the use of fossil fuels. We all know that we are going to either destroy our atmosphere or climate, or run out of viable reserves of oil; the only question is which will happen first! But the easiest course day-to-day is to continue with the current system. The reasons for this inertia range from the power of those special interests that depend on the current system to political lethargy, but such inertia cannot be underestimated. Huge systems require huge forces to divert them from their courses.

---

[12] By this point, it should be obvious that I do not intend to run for political office.

From the perspective of the developed countries, robotics offers a means of filling its needs for these services without resorting to a de facto caste system, or finding other ways to exploit the less fortunate.

It should be our goal as robot designers to lift every soul possible out of such mind-numbing work and to allow society to utilize people at their highest capacity.

Service sector business models

The primary reason that more progress has not been made in service robotics is the lack of capital. Lacking large investment, the small companies in this sector have commonly been forced to bootstrap themselves by building robots in small quantities and attempting to sell them at a profit.

This brings to the discussion the concept of *scalability*. Some very viable markets cannot be scaled down to volumes that allow the bootstrapping method to work. Robots that would cost, only, say $10,000 to build in the thousands may cost $100,000 to build in ones quantities. There may be an insufficient market at this price to allow the bootstrapping process to be started.

Another scalability problem results from customer dispersion. If a start-up company sells only a few robots in a region, it is not practical to station a technician in the area. This means that one of three things must be done to service the robots:

1. Technicians must travel to the robots, greatly increasing service costs.

2. The robots must travel to the service center, greatly increasing downtime and decreasing customer satisfaction.

3. A third party technician must be contracted to handle repairs. This technician will tend to be less well-trained than a factory technician and may have other priorities.

There is another aspect of marketing to the service sector, and that is the fact that services are normally billed by the month. Unlike the customer-base in the manufacturing sector, the customers in this market do not necessarily have capital budgets with which to buy robots. Many executives who handle huge annual budgets for services have never produced a single cost justification for a capital item. So how will this market be won?

The most likely scenario for success in the service industry will be the merging of human and robotic resources into service providers such that potential contracts will be bid in exactly the same way traditional providers bid them. The combination of robots and people will provide the flexibility missing in purely robotic solutions, making these services interchangeable with those of traditional service companies, and the cost savings to the customer will make the choice of a robotic provider a "no brainer." My bet for the breakout is on this concept.

The government sector

In this discussion I will speak to the history of development in the US, even though my experience leads me to believe that many of my observations will be true of other countries as well.

It is widely appreciated that early-stage basic research into revolutionary technologies is seldom undertaken by the private sector because of the huge expenses and risks associated with such investments. Funding by the government is therefore appropriate and essential for a developed country if it is to maintain at least technological parity with other nations.

In recent decades, two examples of US government sponsored research have shown how dramatically such investments can pay off. These are the space program and the Internet. The impact of the space program could never have been imagined when it was begun as a cold war competition with the Soviet Union! Today, satellites provide everything from surveys of crops to weather data and commercial television. From a robotics perspective, the advent of the satellite-based GPS navigation system has meant that autonomous vehicles have become practical in many environments. The impact of the Internet is, if anything, even greater than the space program. The Internet has revolutionized the way we communicate, shop, and search for information, and opened up countless new possibilities for enterprise.

Robotics research and development in the government sector

Here the story is not so encouraging. The US government has made huge investments in autonomous robotics, dating back more than two decades. To say that these investments have failed to meet expectations is a bit like saying that the maiden voyage of the Titanic failed to live up to the promises in the brochure!

To a large extent, these failures are symptomatic of a broken system, and unless it is repaired[13] they will no doubt continue. During the late 1980s and early 1990s, the

---

[13] The likelihood of the system being repaired makes the odds on the lotto look like a slam dunk.

largest programs were sponsored by DARPA, the U.S. Army, the U.S. Air Force and the Department of Energy (DOE).

DARPA

The Defense Advanced Research Projects Agency has the mandate of funding research that is too far ahead of deployment to be undertaken by the various military branches. The strategy was for DARPA to develop technologies to the point where their technical feasibility could be demonstrated and they could be handed over to the other branches for development into operational systems.

The management style of DARPA and the technical competence of its management are very much in contrast with those of the traditional branches of the military.[14] Over the years, they have funded robotics programs through universities, small businesses, and large defense contractors. These programs have succeeded in demonstrating the technical feasibility of a wide-range of autonomous applications. Even so, none of their autonomous robotic programs have successfully transitioned to the various branches of the military.

While the mandate of DARPA is forward-looking research that is not necessarily expected to lead directly to fieldable systems, the expenditures of DOE and the individual branches of the military are expected to be directed toward specific requirements and to culminate in fielded equipment.

Department of Energy

Some of the most interesting and challenging projects with which I am familiar were developed under DOE during the 1980s and early 1990s. These included everything from robotic gantries, to pipe-crawling robots, walking robots, and conventional wheeled and tracked vehicles.

The applications chosen by DOE were usually well considered from the perspective of being appropriate to robotic technology. Applications ran from inspecting welds to removing asbestos insulation from pipes, from radiation surveys of floors to inspecting drums of nuclear waste. Furthermore, the management of the technical aspects of the development phase of these programs was in large part quite professional.

---

[14] I have never personally worked on a DARPA contract, but the reports from companies who have seem to be very positive. This is in stark contrast to my experiences and the reports of my acquaintances regarding other branches.

Prototypes of many robot configurations were developed and tested, often with encouraging results. As the cold war came to an end, these projects focused more on decommissioning and clean-up than producing nuclear weapons. So why were none successfully fielded?

The primary reason for failure in the case of DOE was related to the fact that the research and operational funding were sharply divided. Those funding the research would investigate the needs of the operational groups, then go off and develop solutions. After a few years they would try to get the operational groups to adopt the systems they had developed. The transition was often referred to as "throwing the system over the wall (to the operational groups)." This phrase describes both the process and the problem.

At best, the operational groups viewed the projects as a wasteful nuisance, and at worse, they viewed them as threatening. These groups did not perceive any benefit to themselves if the programs succeeded. This was particularly tragic, because many of the projects were designed primarily to reduce the exposure of personnel to radiation! With few exceptions, the operational groups viewed these projects with disdain and did their best to assure their failure. The failure thus mirrors the problem experienced by DARPA when it tried to pass programs to the military branches.

To the operational groups the very threat of radiation meant that they had much higher pay scales than equivalent workers in the commercial sector. These organizations had attracted people who saw this as a positive opportunity. The two camps became increasingly polarized. The robotics groups, which focused mostly on remediation (clean-up) after 1990, often called the operational attitude a "muck and truck" mentality.

As program after program failed to make it "over the wall," some program managers identified the problem and attempted to get the operational people involved more closely in the development process. Unfortunately, lacking a profit incentive, it was very difficult to accomplish this. In fact, most managers appear to have viewed large staffs as a positive factor in establishing their power and prestige.

In the mid 1990s, new management in DOE put a halt to essentially all robotics work. The "muck and truckers" had won. Dozens of small businesses that had hoped to provide DOE with wondrous new robotic systems suddenly found their one customer gone. With them went all of the knowledge they had accumulated, and the teams they had built.

Flashback...

We participated in the development of ARIES, one of the last big robotics projects funded by DOE in the mid 1990s. The objective was to develop a robot that could travel through a warehouse of low-level "mixed waste." The waste consisted of almost any material that had become contaminated with radioactivity, and it was largely stored in 55 and 85 gallon drums. Various local and federal laws mandated that these drums be inspected frequently to assure they were not leaking. Inspection by humans required being in close proximity to this waste, and this in turn assured that the inspectors would soon receive significant dosages of radiation.

The robot was required to navigate aisles between drums that were a mere 36 inches wide, yet it needed to inspect drums stacked to over 20 feet. In inspecting these drums, the robot would read the drum's barcode, take digital images and perform structured light mapping of the surface of the drums to detect leakage or damage. The results would be compared to data from previous inspections to determine if the drums were deteriorating.

Cybermotion worked with the University of South Carolina and Clemson University to develop the inspection hardware and software. So advanced was the software that it actually built a 3D model of the warehouse as the robot performed its inspection. An operator could fly through this model and immediately spot drums that had been marked in red for attention. Clicking on the model of a drum would bring up its image and history.

The resulting system was installed in a warehouse of the DOE Fernald facility where it was successfully demonstrated. A competing system (IMSS) developed by Lockheed Martin was also demonstrated. Since the Fernald staff had been brought into the program from the beginning, it was expected that they would adopt the system.

Figure 19.1 ARIES Robot at work

Unfortunately, shortly before the operational test it was announced that Fernald would be closed! This resulted in the sponsor offering to demonstrate the system on-site at Los Alamos. We took a team to Los Alamos and installed the system in the bitter cold of late November, 1998. To our relief, the system worked perfectly in the near zero-degree temperatures. The operational staff, however, decided they were not interested and declined to even attend a demonstration. After two weeks of miserable conditions and a significant dose of radiation, we were told to simply pack the system up for shipment to INEEL (Idaho National Engineering and Environmental Laboratory).

At INEEL, instead of setting the system up to inspect real waste, we were directed to set up a small demonstration with empty drums. When we asked what they intended to do with the demonstration, we were told that they often hosted high school class tours, and that it would be a show-and-tell! We were then given a tour of dozens of similar systems, some representing far greater expenditures! All these systems had been "thrown over the wall" with a resounding thud.

U.S. Air Force and Navy

To see any light in DOD's robotic investments, one must look to the skies, and even here the story is clouded at best. During the late 1970s and early 1980s, the U.S. Air Force and Navy began picking up on an idea that seems to have originated with the Israeli military's use of simple remote-controlled aircraft for surveillance. Unsatisfied with what amounted to little more than an oversized model plane, the Air Force began development of far more sophisticated, capable and expensive UAVs (Unmanned Aerial Vehicles).

The initial programs fell well short of expectation, and were widely derided in government circles, but the Air Force and Navy remained undeterred. By the Persian Gulf War in 1990, all services had simple remote-controlled aircraft available that could perform useful surveillance. The Navy's Pioneer UAV, which looked very much like a model of the twin-tailed, WWII era Lightening fighter, acquitted itself well and scored significant marks performing forward observer fire adjustment missions.

The technology began to evolve rapidly, and by the time of the Afghanistan conflict, a UAV called the Predator was capable of not only surveillance, but even direct-attack using Hellfire missiles. But despite the much touted and successful Predator attack on a Taliban concentration south of Kabul, three of its four live-fire missions went wrong, with two of these missions resulting in the deaths of a total of thirteen innocent civilians. Furthermore, half of the twelve Predators deployed during the

2001-2002 period crashed. But on the bright side, these aircraft were still relatively inexpensive and not a single pilot was killed or captured!

The $4.5M Predator's weakness was that it was a half measure, being remotely operated by a two-man crew, and not truly autonomous. It is quite likely that the need for human commands exacerbated the aircraft's problems. Without sufficient built-in intelligence, the aircraft was vulnerable to even moderately bad weather conditions and enemy fire.

One reason for these problems is that teleoperating a vehicle is far more difficult than piloting one[15]. The pilot of a UAV is disconnected from the sensations of flight, and communications delays and interruptions often result in "over control." Furthermore, with a service ceiling of only 10,000 feet, the Predator was very susceptible to ground fire. But viewing the world through a silent camera, it was often difficult for the pilot to even determine that the aircraft was under fire until it was too late. The lesson was obvious; these aircraft needed to be smarter.

The next generation UAV, called the Global Hawk was designed to be just that. Developed in concert with Australia, the Global Hawk promised longer endurance, better sensors, and more autonomy.

In April 2001, a Global Hawk took off from Edwards Air Force Base on the west coast of the US and flew nonstop to RAAF Base Edinburgh, South Australia! Although the two $15M Global Hawks that were rushed to Afghanistan both crashed, the promise of the technology was still attractive to the Air Force.

But these are not the cheap substitutes for manned planes that they were originally expected to be. Between 1995 and 2003, Grumman received $1.6 billion dollars in contracts for the design, development, and testing of the Global Hawk. The most recent contract (Feb. 2003) for four Global Hawks and some support hardware was over $300M! It appears that the Pentagon successfully overcame the looming problem of low cost and managed to bring the cost of UAVs inline with other military expenditures!

On the brighter side, it should also be pointed out that other fully autonomous aircraft have become accepted commodities in the military. Cruise missiles were proving very reliable as early as the Gulf War, and provide the US with the high-tech stand-off weapon of remarkable accuracy. The biggest weakness to the cruise missile (besides its cost of about $.5M), is the fact that it must be targeted before it takes off, and is

---

[15] This is true of ground vehicles and indoor robots as well.

thus not a suitable weapon for targets of opportunity that suddenly appear on the battlefield. Today cruise missiles are being developed that can be deployed to a target area and loiter waiting for further assignment to a precise target, probably by another unmanned targeting aircraft.

U.S. Army

Beginning in the 1980s, the U.S. Army identified a large number of applications that would be ideal for autonomous robots. These included both combat and support roles and ranged from mine-clearing robots to autonomous trucks that could follow a "bread crumb trail" in convoys. Other applications included security patrol both indoors and outdoors, and even an anti-tank role. Configurations ranged from "flying donuts" to autonomous Hum-Vs.

The experiences of the U.S. Army in the occupation of Iraq make it abundantly clear that some of these autonomous systems would probably have saved the lives of many soldiers had they been available in time. *None were.* Some were never funded. Some were funded intermittently, causing the development to repeatedly lose team cohesion. Still others were funded continuously to the cumulative level of hundreds of millions of dollars, but still managed to produce nothing.

The only significant presence of ground robots in Afghanistan and Iraq were tele-operated vehicles. These included the Foster-Miller Talon and Solem, the iRobot PackBot, the Mesa Associates Matilda, and a number of Grumman Remotec robots. These robots were similar to those used for years as EOD (Explosive Ordnance Disposal) devices by various police departments and the FBI (Federal Bureau of Investigation). They were all small- to medium-sized tracked robots and were used largely for exploring caves, for remote surveillance, and in EOD roles.

So where were the autonomous robots for convoy duty and mine clearing? Prototypes had been tested years before. For example, Foster-Miller had developed a prototype autonomous mine-clearing robot by 1993! The answer is tragic.

Designed to fail

The failure of these robotics programs to "be there" for the US troops in Iraq is just another example of the problems afflicting DOD development[16]. Most of these problems are anything but new, but it is important for the entrepreneur to understand the way things really work lest he or she fall victim to the siren's call of chasing the big military contract.[17]

One of the greatest weaknesses of the system is that it is "rule-based." If you will recall our example of a rule-based robot in the first chapter, you will remember that rules were added every time something went wrong. The result quickly became a hopeless mess. This is how the procurement system has evolved. Every time a contractor was caught taking advantage of the government or simply making design mistakes, rules were added. Eventually, the system became so complex that it required an enormous overhead to simply assure compliance with the rules, let alone actually doing anything useful. Nothing could be done inexpensively and yet the mistakes and cheating continued unabated. Thus, the system became negotiable only for companies born in its image and its inefficiencies have become the stuff of legend.

The industry is thus now dominated by the large defense contractors, and they have developed symbiotic relationships with the governmental R&D programs. Despite repeated urgings and legislation from congress, small businesses rarely have an opportunity to become a vendor to any but the smallest procurements. A small company has no chance of breaking into the club, and its only option is to be acquired by one of the big defense contractors. This is not a new situation.

As just one example of this, at the outbreak of WWII, the Defense Department put out a panicked request for someone to design a small battlefield vehicle that was direly needed for the war. The small automotive manufacturer, American Bantam Car Company, that heroically developed the famous WWII Jeep in just six weeks, was then deemed

---

[16] As a lieutenant in Vietnam, I experienced this failure first hand. I was provided with an early version of the M-16 which jammed repeatedly, and with two models of jeep-mounted radios (one FM and one SSB) that were both defective in their design. The radio situation was so bad that we were forced to do completely illegal repairs on the FM (AN-VRC47) radios in order to maintain communications throughout our air cavalry squadron. But this is a story for another book.

[17] One important exception here is the SBIR (Small Business Innovative Research) program. This program provides grants of up to $75k for phase one research, and up to $750k for phase two research. These grants can be extremely important to small companies in funding research and development.

unsuitable as a vendor and was relegated to making trailers to go behind their creations. The contracts for vast numbers of Jeeps were spread between larger automakers.

An even worse problem is that the groups running the development of these programs know that if the project is ever actually fielded, they will be off the gravy train. Since they hire large numbers of contractors (referred to in the industry as beltway bandits) to run the incredible bureaucracy of their programs, there are a lot of salaries at stake.

To justify their salaries, and to assure that the development phase of the program never ends, these consultants regularly think of expensive and usually frivolous new requirements and capabilities that the target system must meet. Each year the systems get more and more expensive.

To add to this, money often flows to other groups in the military to support the programs. This dispersion is done on the basis of almost every criterion except competence. Lacking a profit motive, or even the fear of failure, these groups tend to collect engineers who are comfortable in such an environment. The result is a bloated structure hooked on R&D money, and completely detached from any feelings of responsibility to the taxpayers or soldiers for whom they supposedly work.

Flashback...

The MDARS-I program, which was based on Cybermotion robots, was originally scheduled for deployment in the mid 1990s. In 1995, it was decided that the deployment would have to be delayed in order to convert the base station code just developed by the Navy from the C language (which was becoming obsolete) to the military's ADA language (which was totally inappropriate for such PC-based applications).

Two years later, the completely rewritten code was given two additional years of development to be tested for Y2K compliance. Why a program written after 1995 would ever have been noncompliant, and why a program that made no significant use of the date would need such effort were of course never discussed.

Like the Jeep's creators, Cybermotion was deemed unqualified to make the enhancements to its own system that the Army needed, and the contract was issued instead to General Dynamics. General Dynamics overran the contract several times causing more delays.

Though Cybermotion had provided its standard SR-3 security robots, we had not been a significant participant in the modifications made by the contractor. A few days before a critical test, I was asked by the program manager to find out why the contractor was having so much difficulty with navigation. I did not point out that he had deemed our company technically unqualified for such work, but dutifully went to help. It took me only one day to find and correct the rather minor mistakes being made, but while I was on site something caught my eye.

On a table, was a dazzling array of strange small objects. Some were made of clear plastic, some of stainless steel, and some of wood. They were made in a wide variety of geometric shapes, but obviously intended for some unfathomable purpose. When I asked, I was told that they had been manufactured to the program's specifications to be used in testing the robot's collision avoidance. I asked why so many had been made at such obvious expense. Suppressing a twisted grin, one of the engineers rolled his eyes back and told me these were only random samples and that there were several crates more in storage! I felt I had reached the bottom of the rabbit hole and must soon meet Alice.

Why aren't the robots here yet?

We have taken a brief look at all of the different markets for autonomous robots and yet there are no clear commercial success stories to date, at least by the standards of commerce. Let's examine some of the most common reasons given:

Reason #1 – Technology limitations

Some still believe that existing technology is just too primitive to reliably solve the problems of autonomous robots. This reasoning was probably true a decade ago, but is no longer the case for most practical applications. Certainly an android with human-like capabilities is still a stretch, but simple tasks such as security patrol, materials transport and cleaning are well within the reach of the sensor, computer, and software state of the art today. The only possible challenge here is whether a particular robotic solution can be manufactured to be cost-effective.

Reason #2 – Inflated claims and unrealistic customer expectations

Certainly, this problem existed in the early years of autonomous robot design. Companies often exaggerated or hyped the capabilities of their robots and of robots they expected to produce. Promising exaggerated capabilities for as yet nonexistent products was a common enough phenomenon that engineers coined the term "vaporware" to describe such claims. But potential customers also imagined capabilities that were not existent and had not been claimed or even implied. In fact, I have often been amazed at the powers people have attributed to our robots.

Flashback...

We had just finished installing an early security patrol robot for a large corporate customer. The installation had been particularly challenging because at that time the robot had only sonar navigation, and the walls of the huge lobby of the facility were broken into short sections that stepped in and out every two meters. The robot patrolled this spacious lobby on the ground floor, and rode an elevator to several levels of narrow corridors and offices above.

At night, the air conditioning was turned off, allowing the air quality to approach the alarm level. One night, the robot was patrolling the 3rd floor when it stopped next to the women's restroom and announced an air quality alert. The threat assessment said that the humidity was very high, and when the scanner automatically activated, it detected motion from its microwave. The motion appeared to be coming from inside the restroom. An officer was dispatched but he found nothing in the restroom.

A few hours later the robot repeated the process at the same location on the 2nd floor. For the console operator, this was more than coincidence, so he had the incident investigated again. This time the responding officer put his ear to the wall and heard the sound of running water. As a result, a ruptured water pipe in the floor of the 4th level was eventually found to be the source of both the two incidents and the mysterious lake that had been accumulating in the basement. That this sort of situation might be detected by the robot was never envisioned, and it even surprised us. But the story soon turned against us.

After being repeated amongst female employees, the story evolved from a leaking pipe in the wall to an overflowing sink, and then to an overflowing toilet. The next week, the security manager was confronted by a delegation of female employees insisting that the robot be shut down because it could obviously be used to see them in the toilet stalls! But the story does not end there...

We had just finished clearing up this misunderstanding and making things safe for the robot once again. The security manager and I had been laughing about the overestimation of the robot's capabilities by these employees. As I was preparing to leave for my plane, he said, "By the way, the next time you are out here we would like you to program your robots to follow the cleaning staff out to the dumpsters and to tell us if they are throwing any stolen items in them."[18]

---

[18] Sometimes called staging, this trick is a common way that employees are known to steal from their employers. The items are thrown away using identifiable trash bags, and the garbage collectors working with the employees retrieve them after picking up the trash.

Reason #3 — A lack of resources

As we have already discussed, this argument obviously has merit, at least in the West. Not only have most small companies failed to obtain the resources required to penetrate their markets, but the larger companies who have these resources have not invested them in the robotics sector for more than a few years at a time.

Figure 19.2. ASIMO and P2 Robots
(Photo courtesy of Honda Motor Co. Ltd.)

The exception would appear to be Japan, where medium and large corporations alike have made patient, long-term investments in autonomous robots. Corporate giant Honda has put huge investments into ASIMO (Advanced Step in Innovative Mobility[19]). A walking android, ASIMO is the result of a corporate research program that began in 1986. It has two degrees of freedom in its neck, six on each arm and six on each leg. It balances and walks, even up stairs. ASIMO's hands have four fingers and a thumb, and it can grasp small objects. In all, ASIMO has 26 motors.

The extent of its autonomy is difficult to ascertain from Honda's publicity, yet as a mechanical marvel it is clearly unparalleled. And Honda funds the development year after year with no clear profit or marketing objective. According to Honda's web site:

> *ASIMO was created to be a helper to people in need. ASIMO's people-friendly size of 4 feet tall (120 centimeters) makes it the perfect size for helping around the house, especially for a person confined to a bed or a wheelchair. ASIMO's size also allows it to look directly at an adult sitting in a chair or sitting up in bed.*

---

[19] The name is obviously designed to pay tribute to Isaac Asimov, whose science fiction writings have featured robots dedicated to the service of people and have inspired the industry. Asimov's three laws of robotics have become better known among robot enthusiasts than most real laws.

Sogo Keibi Hosho, another Japanese company has spent more than a decade developing its Guard Robot C4, with no significant sales. This type of faith would seem foolhardy to most western managers, but it shows the absolute faith that the Japanese have in the future of robots in our lives.

Reason #4 – Acceptance, fear and prejudice

In my opinion, the most important single obstacle to success is public acceptance and more precisely customer acceptance. While people are fascinated by robots, they are also threatened. I have tethered robots through countless buildings on their way to demonstrations, and the most common first reaction I have heard from the people in those places has not been "What does it do?" or "How much does it cost?" The most common first reaction has been to ask rhetorically "Is that thing going to take my job?"

Instead of seeing the potential of robots to make their lives better, people more often conjure up their darkest fears. It took more than a decade for labor in the American automotive manufacturing sector to even begin to understand that automation was their only chance of *keeping* their jobs in a world economy where labor abroad is a tiny fraction of that at home.

In my opinion, this issue will eventually decide the matter. At some point, the common perception of robots will shift from the negative to the positive, and all other factors will change in response.

The future

Given this lengthy diatribe about failings and disappointments, one might expect that my opinion of the future of autonomous robotics would be bleak. Nothing could be further from the truth. Autonomous robots will have their time, and nothing short of global catastrophe will stop it. In this, I am absolutely sure.

Far from being the result of blind faith or wishful thinking, this opinion is based on the inextricable conclusions of geometric analysis. Virtually every factor bearing on the cost and capability of robots, and even factors bearing on the need for them, are all moving in favor of their eventual acceptance.

Enabling technologies

Enabling technologies ranging from sensors to radio communications and navigation aids are all accelerating logarithmically. The ubiquitous acceptance of wireless LAN systems, the plunging costs of video cameras and processors, the availability of affordable laser navigation systems, and the ever-increasing accuracy and dropping cost of GPS navigation receivers are all combining to make autonomous robots potentially cheaper and ever more capable.

At least as important, we now have enormous resources in human experience. Countless software engineers and academics have spent endless hours developing concepts of modeling and control that are just as much part of the existing robotics toolbox as any sensor or processor. As a result, only the integration of these elements is required for new robotic configurations to burst onto the scene with blinding speed.

Social forces

The social issues already discussed are pushing customers to look for new solutions to performing many of the tasks that now require manual labor. These are tasks which autonomous robots can easily provide. Slowly but surely, a few venture capitalists (real ones) are beginning to make investments in companies like iRobot, and the industry is beginning to gain a little attention.[20]

Timing

Waiting for the robot revolution is a bit like waiting to see which straw will break the proverbial camel's back. When it happens, it will happen quickly. Competitors will appear, products will improve rapidly, and prices will fall in response to the savings from mass production. Falling prices and growing capabilities will further spur the market to grow. But many factors will influence the exact timing. To understand the likely scenario, it is useful to look at other technologies and their acceptance. We have already discussed the long road to acceptance that industrial robots traveled.

Microwave ovens were developed during WWII, and video recorders were invented shortly thereafter. The microwave oven would have to wait for women to reenter[21]

---

[20] I should caution that there have been several false starts in the past, and this may be yet another.

[21] Many women in the US had returned to being housewives following WWII, but they had proven what they could do in the war industries and did not stay home long.

the workplace in the late 1960s. As they did, they found themselves with no time or energy to cook elaborate meals over an open stove, and the market for prepackaged foods and microwave ovens took off.

Reel-to-reel video recorders were available starting in the 1950s, but they achieved little penetration of the consumer market. This lack of a market kept the costs high and the market low. Sony reasoned that less expensive players, and the availability of rental movies on video cassettes would jump-start the market, but their attempts to make titles available did not prove decisive. Only when they included a timer to allow "time-shifting" for viewers did the video recorder begin to have its day in the sun. That, after all Sony's efforts, its Beta technology should lose out to the arguably inferior VHS standard is another of the many ironies that permeate this avalanche-like process.

Personal computers languished for years before IBM entered the market. The mere entrance of this icon of the computer world convinced the market that the PC was a real and indispensable tool. Bill Gate's appreciation for the importance of the moment, and the imminent shift of profit centers from hardware to software, allowed Microsoft to become a world economic power while IBM gained only marginally from precipitating the revolution.

These examples of emerging technologies go on and on. In every case, the technology was available for a decade or more before the acceptance conditions were reached. Autonomous robots have served their time in purgatory, and now the countdown to their time has begun. Will the Japanese investment in autonomous robots pay off as it did in industrial robots, or will a twist of fate rob them of their prize? Whatever happens, make no mistake, the day of the autonomous robot is at hand.

APPENDIX:
Referenced Laws and Formulas

Law of Sines and Law of Cosines

The law of sines is used to find angles of a general triangle. If two sides and the enclosed angle are known, they can be used in conjunction with the law of sines to find the third side and the other two angles.

The law of cosines is used for calculating one side of a triangle when the angle opposite and the other two sides are known. It can be used in conjunction with the law of sines to find all sides and angles.

If two interior angles are known, the third can be calculated from the fact that the sum of the interior angles is always 180 degrees.

Law of Sines:

$$\frac{a}{Sin\ A} = \frac{b}{Sin\ B} = \frac{c}{Sin\ C}$$

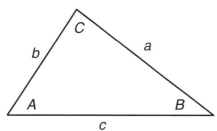

Law of Cosines:

$$c^2 = a^2 + b^2 - 2ab\ Cos\ C$$

$$A + B + C = 180\ \text{degrees}$$

Simple 2D Vector Addition
(Using map heading convention)

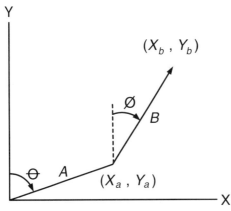

$$Y_a = A \cos \Theta$$

$$X_a = A \sin \Theta$$

$$Y_b = (A \cos \Theta) + (B \cos \emptyset)$$

$$X_b = (A \sin \Theta) + (B \sin \emptyset)$$

Linear Regression

Linear regression can be used to fit the line

$y = mx + b$ to linear data where:

1. x is the dependent variable
2. y is the independent variable
3. x_i is the x value for ith data point
4. y_i is the y value for the ith data point
5. N is the number of different standards used
6. y_{ave} is the average of the y values for the standards
7. x_{ave} is the average of the x values for the standards

This method assumes that there is no variance in the value for x and that each standard is analyzed once.

Calculate Sums:

$$S_{xy} = ss_{xy} = \sum (x - x_{ave}) \cdot (y - y_{ave}) \text{ or } \sum (x_i \cdot y_i) - \left(\frac{\sum x_i \cdot \sum y_i}{N} \right)$$

$$S_{xy} = ss_{xy} = \sum (x - x_{ave}) \cdot (y - y_{ave}) \text{ or } \sum (x_i \cdot y_i) - \left(\frac{\sum x_i \cdot \sum y_i}{N} \right)$$

$$S_{yy} = ss_{yy} = \sum (y - y_{ave})^2 \text{ or } \sum (y_i)^2 - \left[\frac{\sum (y_i)^2}{N} \right]$$

Calculating Slope and Intercept:

$$\text{Slope} = m = \frac{S_{xy}}{S_{xx}}$$

$$\text{Intercept} = b = y_{ave} - (m \cdot x_{ave})$$

Uncertainty (standard deviation) in Regression:
(Assuming linear function and no replicates)

$$s_r = \sqrt{\frac{s_{yy} - \left(m^2 \cdot s_{xx}\right)}{\left(N-2\right)}}$$

Uncertainty in $y_{prediction}$:

$$s_y = s_r \cdot \sqrt{1 + \left(\frac{1}{N}\right) + \frac{\left(x - x_{ave}\right)^2}{s_{xx}}}$$

Uncertainty in $x_{prediction}$:

$$s_x = \left(\frac{s_r}{m}\right) \cdot \sqrt{\left(\frac{1}{M}\right) + \left(\frac{1}{N}\right) + \left[\frac{\left(y_{unk} - y_{ave}\right)^2}{m^2 \cdot s_{xx}}\right]}$$

About the Author

John M. Holland is a well-known pioneer of mobile robotics, founding Cybermotion in 1984, the first company to successfully manufacture and sell commercial mobile robot units, to customers such as the U.S. Department of Energy, U.S. Army, Boeing, NASA, General Motors and many others. An electrical engineer by training, John holds six U.S. patents and is the author of two previous books, including the foundational book *Basic Robotics Concepts* (Howard Sams, 1983). He has written and lectured extensively and is an irreverent and outspoken futurist.

Index

ELSEVIER SCIENCE CD-ROM LICENSE AGREEMENT

PLEASE READ THE FOLLOWING AGREEMENT CAREFULLY BEFORE USING THIS CD-ROM PRODUCT. THIS CD-ROM PRODUCT IS LICENSED UNDER THE TERMS CONTAINED IN THIS CD-ROM LICENSE AGREEMENT ("Agreement"). BY USING THIS CD-ROM PRODUCT, YOU, AN INDIVIDUAL OR ENTITY INCLUDING EMPLOYEES, AGENTS AND REPRESENTATIVES ("You" or "Your"), ACKNOWLEDGE THAT YOU HAVE READ THIS AGREEMENT, THAT YOU UNDERSTAND IT, AND THAT YOU AGREE TO BE BOUND BY THE TERMS AND CONDITIONS OF THIS AGREEMENT. ELSEVIER SCIENCE INC. ("Elsevier Science") EXPRESSLY DOES NOT AGREE TO LICENSE THIS CD-ROM PRODUCT TO YOU UNLESS YOU ASSENT TO THIS AGREEMENT. IF YOU DO NOT AGREE WITH ANY OF THE FOLLOWING TERMS, YOU MAY, WITHIN THIRTY (30) DAYS AFTER YOUR RECEIPT OF THIS CD-ROM PRODUCT RETURN THE UNUSED CD-ROM PRODUCT AND ALL ACCOMPANYING DOCUMENTATION TO ELSEVIER SCIENCE FOR A FULL REFUND.

DEFINITIONS

As used in this Agreement, these terms shall have the following meanings:

"Proprietary Material" means the valuable and proprietary information content of this CD-ROM Product including all indexes and graphic materials and software used to access, index, search and retrieve the information content from this CD-ROM Product developed or licensed by Elsevier Science and/or its affiliates, suppliers and licensors.

"CD-ROM Product" means the copy of the Proprietary Material and any other material delivered on CD-ROM and any other human-readable or machine-readable materials enclosed with this Agreement, including without limitation documentation relating to the same.

OWNERSHIP

This CD-ROM Product has been supplied by and is proprietary to Elsevier Science and/or its affiliates, suppliers and licensors. The copyright in the CD-ROM Product belongs to Elsevier Science and/or its affiliates, suppliers and licensors and is protected by the national and state copyright, trademark, trade secret and other intellectual property laws of the United States and international treaty provisions, including without limitation the Universal Copyright Convention and the Berne Copyright Convention. You have no ownership rights in this CD-ROM Product. Except as expressly set forth herein, no part of this CD-ROM Product, including without limitation the Proprietary Material, may be modified, copied or distributed in hardcopy or machine-readable form without prior written consent from Elsevier Science. All rights not expressly granted to You herein are expressly reserved. Any other use of this CD-ROM Product by any person or entity is strictly prohibited and a violation of this Agreement.

SCOPE OF RIGHTS LICENSED (PERMITTED USES)

Elsevier Science is granting to You a limited, non-exclusive, non-transferable license to use this CD-ROM Product in accordance with the terms of this Agreement. You may use or provide access to this CD-ROM Product on a single computer or terminal physically located at Your premises and in a secure network or move this CD-ROM Product to and use it on another single computer or terminal at the same location for personal use only, but under no circumstances may You use or provide access to any part or parts of this CD-ROM Product on more than one computer or terminal simultaneously.

You shall not (a) copy, download, or otherwise reproduce the CD-ROM Product in any medium, including, without limitation, online transmissions, local area networks, wide area networks, intranets, extranets and the Internet, or in any way, in whole or in part, except that You may print or download limited portions of the Proprietary Material that are the results of discrete searches; (b) alter, modify, or adapt the CD-ROM Product, including but not limited to decompiling, disassembling, reverse engineering, or creating derivative works, without the prior written approval of Elsevier Science; (c) sell, license or otherwise distribute to third parties the CD-ROM Product or any part or parts thereof; or (d) alter, remove, obscure or obstruct the display of any copyright, trademark or other proprietary notice on or in the CD-ROM Product or on any printout or download of portions of the Proprietary Materials.

RESTRICTIONS ON TRANSFER

This License is personal to You, and neither Your rights hereunder nor the tangible embodiments of this CD-ROM Product, including without limitation the Proprietary Material, may be sold, assigned, transferred or sub-licensed to any other person, including without limitation by operation of law, without the prior written consent of Elsevier Science. Any purported sale, assignment, transfer or sublicense without the prior written consent of Elsevier Science will be void and will automatically terminate the License granted hereunder.

TERM

This Agreement will remain in effect until terminated pursuant to the terms of this Agreement. You may terminate this Agreement at any time by removing from Your system and destroying the CD-ROM Product. Unauthorized copying of the CD-ROM Product, including without limitation, the Proprietary Material and documentation, or otherwise failing to comply with the terms and conditions of this Agreement shall result in automatic termination of this license and will make available to Elsevier Science legal remedies. Upon termination of this Agreement, the license granted herein will terminate and You must immediately destroy the CD-ROM Product and accompanying documentation. All provisions relating to proprietary rights shall survive termination of this Agreement.

LIMITED WARRANTY AND LIMITATION OF LIABILITY

NEITHER ELSEVIER SCIENCE NOR ITS LICENSORS REPRESENT OR WARRANT THAT THE INFORMATION CON-TAINED IN THE PROPRIETARY MATERIALS IS COMPLETE OR FREE FROM ERROR, AND NEITHER ASSUMES, AND BOTH EXPRESSLY DISCLAIM, ANY LIABILITY TO ANY PERSON FOR ANY LOSS OR DAMAGE CAUSED BY ERRORS OR OMISSIONS IN THE PROPRIETARY MATERIAL, WHETHER SUCH ERRORS OR OMISSIONS RESULT FROM NEG-LIGENCE, ACCIDENT, OR ANY OTHER CAUSE. IN ADDITION, NEITHER ELSEVIER SCIENCE NOR ITS LICENSORS MAKE ANY REPRESENTATIONS OR WARRANTIES, EITHER EXPRESS OR IMPLIED, REGARDING THE PERFOR-MANCE OF YOUR NETWORK OR COMPUTER SYSTEM WHEN USED IN CONJUNCTION WITH THE CD-ROM PRODUCT.

If this CD-ROM Product is defective, Elsevier Science will replace it at no charge if the defective CD-ROM Product is returned to Elsevier Science within sixty (60) days (or the greatest period allowable by applicable law) from the date of shipment.

Elsevier Science warrants that the software embodied in this CD-ROM Product will perform in substantial compliance with the documentation supplied in this CD-ROM Product. If You report significant defect in performance in writing to Elsevier Science, and Elsevier Science is not able to correct same within sixty (60) days after its receipt of Your notification, You may return this CD-ROM Product, including all copies and documentation, to Elsevier Science and Elsevier Science will refund Your money.

YOU UNDERSTAND THAT, EXCEPT FOR THE 60-DAY LIMITED WARRANTY RECITED ABOVE, ELSEVIER SCIENCE, ITS AFFILIATES, LICENSORS, SUPPLIERS AND AGENTS, MAKE NO WARRANTIES, EXPRESSED OR IMPLIED, WITH RESPECT TO THE CD-ROM PRODUCT, INCLUDING, WITHOUT LIMITATION THE PROPRIETARY MATERIAL, AN SPECIFICALLY DISCLAIM ANY WARRANTY OF MERCHANTABILITY OR FITNESS FOR A PARTICULAR PURPOSE.

If the information provided on this CD-ROM contains medical or health sciences information, it is intended for professional use within the medical field. Information about medical treatment or drug dosages is intended strictly for professional use, and because of rapid advances in the medical sciences, independent verification f diagnosis and drug dosages should be made.

IN NO EVENT WILL ELSEVIER SCIENCE, ITS AFFILIATES, LICENSORS, SUPPLIERS OR AGENTS, BE LIABLE TO YOU FOR ANY DAMAGES, INCLUDING, WITHOUT LIMITATION, ANY LOST PROFITS, LOST SAVINGS OR OTHER IN-CIDENTAL OR CONSEQUENTIAL DAMAGES, ARISING OUT OF YOUR USE OR INABILITY TO USE THE CD-ROM PRODUCT REGARDLESS OF WHETHER SUCH DAMAGES ARE FORESEEABLE OR WHETHER SUCH DAMAGES ARE DEEMED TO RESULT FROM THE FAILURE OR INADEQUACY OF ANY EXCLUSIVE OR OTHER REMEDY.

U.S. GOVERNMENT RESTRICTED RIGHTS

The CD-ROM Product and documentation are provided with restricted rights. Use, duplication or disclosure by the U.S. Govern-ment is subject to restrictions as set forth in subparagraphs (a) through (d) of the Commercial Computer Restricted Rights clause at FAR 52.22719 or in subparagraph (c)(1)(ii) of the Rights in Technical Data and Computer Software clause at DFARS 252.2277013, or at 252.2117015, as applicable. Contractor/Manufacturer is Elsevier Science Inc., 655 Avenue of the Americas, New York, NY 10010-5107 USA.

GOVERNING LAW

This Agreement shall be governed by the laws of the State of New York, USA. In any dispute arising out of this Agreement, you and Elsevier Science each consent to the exclusive personal jurisdiction and venue in the state and federal courts within New York County, New York, USA.